Lasers and Electro-Optics Research and Technology Series

FROM FEMTO-TO ATTOSCIENCE AND BEYOND

Lasers and Electro-Optics Research and Technology Series

High-Power and Femtosecond Lasers: Properties, Materials and Applications
Paul-Henri Barret and Michael Palmer (Editors)
2009. ISBN: 978-1-60741-009-6

Fiber Lasers: Research, Technology and Applications
Masato Kimura (Editor)
2009. ISBN: 978-1-60692-896-7

Optical Solitons in Nonlinear Micro Ring Resonators: Unexpected Results and Applications
Nithiroth Pornsuwancharoen, Jalil Ali andPreecha Yupapin
2009. ISBN: 978-1-60741-342-4

From Femto-to Attascience and Beyond
Janina Marciak-Kozlowska and Miroslaw Kozlowski
2009. ISBN: 978-1-60741-164-2

Lasers and Electro-Optics Research and Technology Series

From Femto-to Attoscience and Beyond

Janina Marciak-Kozlowska
and
Miroslaw Kozlowski

Nova Science Publishers, Inc.
New York

LIBRARY OF CONGRESS CATALOGING-IN-PUBLICATION DATA

Marciak-Kozlowska, Janina.
From femto-to attoscience and beyond / Janina Marciak-Kozlowska and Miroslaw Kozlowski.
p. cm.
Includes index.
ISBN 978-1-60741-164-2 (hardcover)
1. Laser pulses, Ultrashort. 2. Rydberg states. 3. Transport theory--Mathematics. I. Kozlowski, Miroslaw. II. Title.
QC689.5.L37M37 2009
621.36'6--dc22

2009004757

Published by Nova Science Publishers, Inc. ✣ New York

CONTENTS

PREFACE

Fascinating developments in optical pulse engineering over the last 20 years lead to the generation of laser pulses as short as few femtosecond, providing a unique tool for high resolution time domain spectroscopy. However, a number of the processes in nature evolve with characteristic times of the order of 1 fs or even shorter. Time domain studies of such processes require at first place sub-fs resolution, offered by pulse depicting attosecond localization. The generation, characterization and proof of principle applications of such pulses is the target of the attoscience.

In this book the interaction of the attosecond laser pulses with matter is investigated. The Proca's termal equation for laster - matter interaction is formulated and solved. It is shown that for attosecond laster pulses the Proca's thermal equation can be simplified and thermal Klein-Gordon equation is obtained.

The book is divided into six chapters and the Epilogue presents the contemporary status of the attoscience: sub-femtosecond laster-matter interactions, master equations: Proca and Klein-Gordon equations and the perspectives: Attosecond Free Electron Laser (AFEL), LASETRON and Entangled Photon Laser (EPL).

INTRODUCTION

The increase in light intensity available in the laboratory over the previous 20 years has been astounding. Laser peak power has climbed from giga-watts to petawatts in this time span, and accessible focused intensity has increased by at least seven orders of magnitude. Such a dramatic increase in light brightness has accessed an entirely new set of phenomena. High repetition rate table top lasers can routinely produce intensity in excess of 10^{19} W/cm^2, and intensities of up to 10^{20} W/cm^2 are possible with the latest petawatt class systems [1]. Light-matter interactions with single atoms are strongly non-perturbative and electron energies are relativistic. The intrinsic energy density of these focused pulses is very high, exceeding a gigajoule per cm^3. The interactions of such intense light with matter lead to dramatic effects, such as high temperature plasma creation, bright X-ray pulse generation, fusion plasma production, relativistic particle acceleration, and highly charged ion production.

Such exotic laser-matter interactions have led to an interesting set of applications in high field science, and high energy density physics (HED physics). These applications span basic science and extend into unexpected new areas such as fusion energy development and astrophysics. In this paper some of these new applications will be reviewed. The topics covered here do not represent a comprehensive list of applications made possible with high intensity short pulse lasers, but they do give a flavor of the diverse areas affected by the latest laser technology. Most of the applications discussed here are based on recent experiments using lasers with peak power of 5-100 TW. Many important leaps in laser technology have driven the rapid advances in HED and high field science over the last 15 years. The enabling advancement for this technological progress was the invention of chirped pulse amplification (CPA) lasers [2]. A broad bandwidth, mode-locked laser produces a low power, ultrafast pulse of light, usually with duration of 20-500 fs. This short pulse is first stretched in time by a factor of around ten thousand from its original duration using diffraction gratings. This allows the pulses, now of much lower peak power, to be safely amplified in the laser, avoiding the deleterious nonlinear effects which would occur if the pulses had higher peak power (Koechner, 1996). These amplified pulses are, finally, recompressed in time (again using gratings), in a manner that preserves the phase relationship between the component frequencies in the pulse. The CPA output has a duration near that of the original pulse but with an energy greater by the amplification factor. In high-energy CPA systems, severe nonlinearities occurring when the pulse propagates in air can be a major problem, so the pulse must be recompressed in an evacuated chamber.

The first generation of CPA lasers were based mainly on flashlamp pumped Nd:glass amplifiers [2]. These glass based lasers are usually limited to pulse duration of greater than about 300 fs because of gain narrowing in the amplifiers (Blanchot et al., 1995). The most significant scaling of this approach to CPA was demonstrated by the Petawatt laser at Lawrence Livermore National Laboratory in the late 1990s. This laser demonstrated the production of 500 J per pulse energy with duration of under 500 fs yielding over 10^{15} W of peak power. Since this demonstration, a number of petawatt laser projects have been undertaken around the world.

The second common approach to CPA uses Ti:sapphire as the amplifier material. This material permits amplification of much shorter pulse durations often down to 30 fs. However, the short excited state lifetime of Ti:sapphire (~3us) requires that the material be pumped by a second laser (usually a frequency doubled Nd:YAG or Nd:glass laser). The inherent inefficiencies of those two step pumping usually limit the output energy of such a laser to under a few joules of energy per pulse. A number of multi-terawatt lasers based on Ti:sapphire now operate in many high intensity laser labs world wide.

The third major technology upon which a new generation of high peak power CPA lasers is being based is optical parametric chirped pulse amplification (OPCPA). In this approach, amplification of the stretched pulses occurs not with an energy storage medium like Nd:glass or Ti:sapphire but via parametric interactions in a nonlinear crystal. This approach is quite attractive because of the very high gain per stage possible (often in excess of 104 per pass) and the very broad gain bandwidth possible in principal. To date, a number of CPA demonstrations with OPCPA have been published. Though there are still significant technological issues to be resolved with OPCPA, this technique promises to lead to a new generation of high peak power, femtosecond lasers.

The physics accessible with this class of lasers is quite extreme. The science applications made possible with these extremes can be simply classified into two categories. First, by temporally compressing the pulses and focusing to spots of a few wavelengths in diameter, these lasers concentrate energy in a very small volume. A multi-TW laser focused to a few microns has an intrinsic energy density of over 10^9 J/cm^3. This corresponds to about l0 keV of energy per atom in a material at solid density. As a result, quite high temperatures can be obtained. Such energy density corresponds to pressure in excess of l0 Gbar. Many applications of ultraintense lasers stem from the ability to concentrate energy to high energy-density which can lead to quite extreme states of matter.

The second class of applications arises from the high field strengths associated with a very intense laser pulse. At an intensity of over 10^{18} W/cm^2, a intensity quite easily achievable with modern table top terawatt lasers, the electric field of the laser exceeds ten atomic units, over 10^{10} V/cm. Consequently, the strong field rapidly ionizes the atoms and molecules during a few laser cycles. At intensity approaching 10^{21} W/cm^2, the current state of the art with petawatt class lasers, the electric field is comparable to the field felt by a K-shell electron in mid Z elements such as argon. The magnetic field is nearly 1000 T. An electron quivering in such a strong electric field will be accelerated to many MeV of energy in a single optical cycle. Such charged particles will experience a very strong forward directed force resulting from the Lorentz force.

High peak power femtosecond lasers are unique in their ability to concentrate energy in a small volume. A dramatic consequence of this concentration of energy is the ability to create matter at high temperature and pressure. Matter with temperature and density near the center of dense stars can be created in the laboratory with the latest high intensity lasers. For example, solid density matter can be heated to temperature of over keV. Under these conditions, the particle pressure inside the sample is over 1 billion atmospheres, far higher than any other pressure found naturally on or in the earth and approaches pressures created in nuclear weapons and inertial confinement fusion implosions.

Study of the properties of matter at these extreme conditions, namely higher in the temperature range of 1-1000 eV, is crucial to understanding many diverse phenomenon, such as the structure of planetary and stellar interiors or how controlled nuclear fusion implosions (inertial confinement fusion or ICF) evolve. Yet, despite the wide technological and astrophysical applications, a true, complete understanding of matter in this regime is not in hand. A large obstacle is posed by the fact that theoretical models of this kind of matter are difficult to formulate. While the atoms in these warm and hot dense plasmas are strongly ionized, the very strong coupling of the plasma, and continuum lowering in the plasma dramatically complicates traditional plasma models which depend on two body collision kinetics. Even the question of whether electrons in this state are free or bound is not as clear cut as it is in a diffuse plasma.

Over the last 15 years, since the development of high energy short pulse lasers, there have been a number of experimental studies aimed at isochorically heating solid density targets with a femtosecond laser pulse (Audebert et al., 2002; Widmann et al., 2001). This class of experiments uses a femtosecond pulse to heat, at solid density, an inertially confined target on a time scale much faster than the hot material can expand by hydrodynamic pressure. This approach can be very powerful. Not only can spectroscopic diagnostics be implemented to derive information on the heated material (such as ionization state) (Nantel et al., 1998) but isochoric heating experiments can also enable laser heated pump-probe experiments. Pump-probe experiments can use an optical, or laser-generated X-ray pulse to probe the material on a time scale before it can expand. This yields, for example, conductivity information through reflectivity and transmission probing (Widmann et al., 2001), as well as XUV and X-ray opacity if a short wavelength probe is employed (Workman et al., 1997). The laser heats a thin slab of material, with thickness usually comparable to or less than an optical skin depth. When the material is heated from its initially cold, solid state, the plasma will be in the "strongly coupled" regime1, even at temperature approaching 1 keV.

Direct laser heating has drawbacks. Because of the small depth of such a skin depth, the expansion time of the sample is often only 100 or 200 fs. The few hundred fs release time of the heated material is comparable to the electron-ion equilibration time and hampers interpretation of this kind of experiment. Microsopically rough surface finish of the very thin target often leads to a large uncertainty in the initial density, complicating analysis of the data.

The plasma particle accelerators have received a considerable interest in the past decade. Improvements on the laser technology and laboratory facilities made the plasma particle accelerators a strong alternative for the huge high-energy colliders. One sort of the plasma particle accelerators is called as the laser wakefield accelerators (LWFA), in which a short, intense laser pulse is used to create a wake in the plasma. Current systems of high power laser pulses produce a plasma wave which can accelerate elections to the energies of one trillion electron volts in a distance less man a centimeter.

When a short, intense laser pulse is sent to plasma, it generates a plasma wave whose amplitude is larger man mat of the wave produced by the single laser pulse using the same total energy. Another series of pulses is used to push extra electrons into the wave, which accelerates them into a narrowly focused beam at nearly the speed of light However, the problem is basically interaction of electromagnetic wave with particles, properties of the medium in which the particles present has great importance. The limits on the LWFA due to properties of the plasma are discussed.

The plasma fluctuations while the laser pulse propagates inside are described by the fluid equations. Assuming that the plasma is cold, and there is no background magnetic field, the motion of electrons, taking the background ions are stationary, is described by

$$\frac{\partial p}{\partial t} + \vec{V}\cdot\vec{\nabla}_p + \nabla P = -e\left[\vec{E} + \left(\frac{\vec{V}}{c}\times\vec{B}\right)\right], \tag{1}$$

where υ and n are the velocity and the density of die electrons, respectively, P is the pressure. Furthermore, the equation of continuity and the Poisson's equation are employed.

$$\frac{\partial n}{\partial t} + \nabla\cdot(n\vec{V}) = 0, \tag{2}$$

$$\nabla\vec{E} = 4\pi\rho. \tag{3}$$

After defining the momentum p, as the product of mass, velocity, and the number density, linearization of the velocity and density together with taking the ponderomotive force into consideration yields

$$m\frac{\partial V}{\partial t} + mv_{ei}V = -e(\nabla\Phi + \nabla\Phi_{NL}). \tag{4}$$

Substituting the Poisson's equation into Eq. (4), a second-order partial differential equation for the density is obtained:

$$\frac{\partial^2 n}{\partial t^2} + \upsilon_{ei}\frac{\partial n}{\partial t} + n\omega_p^2 = -\frac{\omega_p^2}{2\pi e}\nabla^2\Phi_{NL}, \tag{5}$$

where ω_p is the plasma frequency, υ_{ei} is the electron ion collision frequency, and Φ_{NL} is the nonlinear potential for the ponderomotive force of the laser beam. The Poisson's equation is used once more to define the number density in terms of the electrostatic potential. Having introduced this fact, the fluctuations in the plasma will be totally described by an inhomogeneous differential equation for the electrostatic potential. The nonlinear potential is well defined in terms of the normalized vector potential $\vec{a}$ such as $\Phi_{NL} = -\left(\frac{mc^2}{2e}\right)\left|\vec{a}^2(r,z,t)\right|$. Assuming the laser beam profile as bi-Gaussian and inducing the degrees of freedom of the problem by combining the time dependency and the longitudinal displacement together such that defining a time-dependent displacement $\xi = z - \upsilon_p t$, where υ_p is the phase velocity of the exerted plasma wave, the nonlinear potential is then described as

$$\Phi_{NL} = -\frac{mc^2}{2e} a_0^2 e^{-\left[-\left(2r^2/\sigma_r^2\right)-\left(\xi^2/\sigma_z^2\right)\right]}. \tag{6}$$

Here the parameters σ_z and σ_r are rms pulse length and spot size, respectively. Thus, Eq. (5) turns out to be

$$\frac{\partial^2\Phi(r,\xi)}{\partial y^2} + 2\alpha\frac{\partial\Phi(r,\xi)}{\partial\xi} + k_p^2\Phi(r,\xi) = -\frac{mk_p^2c^2}{2e} a_0^2 e^{-\left[\left(2r^2/\sigma_r^2\right)+\left(\xi^2/\sigma_z^2\right)\right]}, \tag{7}$$

where $a = \frac{\upsilon_{e_i}}{2\upsilon_p}$, and $k_p = \frac{\omega_p}{\upsilon_p}$. Making use of the error functions and the limits $\xi \to -\infty$ the solution of Eq. (7) for electric fields (wakefields) is expressed as

$$E_z(r,\xi) = \frac{\sqrt{\pi}}{4e} mc^2\sigma_z k_p^2 e^{-\left[-\left(k^2\sigma_r^2/4\right)-\left(r^2\sigma_r^2\right)\right]} \cdot \cos k\xi, \tag{8}$$

$$E_r(r,\xi) = -\sqrt{\pi}\frac{mc^2}{e} a_0^2 \frac{r}{\sigma_r^2}\sigma_z e^{-\left(2r^2/\sigma_r^2\right)+\left(k_p^2\sigma_z^2\right)} \cdot \sin k\xi. \tag{9}$$

The axial wakefield reaches the maximum amplitude at $\xi = \frac{n\pi}{k}, (n = 0, 1, 2,\ldots)$. Naturally, these extremes alternate depending on the integer n, such that E_z has a maximum for even n and it has a minimum for odd n on the z axis. More clearly it has a maximum at $z = \upsilon_p t,\ \upsilon_p t + \frac{2\pi}{k}, \upsilon_p t + \frac{4\pi}{k},\ldots$ and it has a minimum at points $z = \upsilon_p t + \frac{\pi}{k}, \upsilon_p t + \frac{3\pi}{k},\ldots$

When the amplitude of the laser beam approaches the critical electric field of quantum electrodynamics (Schwinger field) the vacuum becomes polarized and electron – positron pairs are created in vacuum. On a distance equal to the Compton length, $\lambda\!\!\!^{-}_C = \hbar/m_e c$, the work of critical field on an electron is equal to the electron rest mass energy $m_e c^2$, i.e. $eE_{Sch}\lambda\!\!\!^{-}_C = m_e c^2$. The dimensionless parameter

$$\frac{E}{E_{Sch}} = \frac{e\hbar E}{m_e^2 c^3} \tag{11}$$

becomes equal to unity for electromagnetic wave intensity of the order of

$$I = \frac{c}{r_e \lambda_C^2} \frac{m_e c^2}{4\pi} \cong 4.7 \cdot 10^{29} \tfrac{W}{cm^2}, \tag{12}$$

where r_e is the classical electron radius. For such ultra high intensities the effects of nonlinear quantum electrodynamics plays a key role: laser beams excite virtual electron – positron pairs. As a result the vacuum acquires a finite electric and magnetic susceptibility which lead to the scattering of light by light. The cross section for the photon – photon interaction is given by:

$$\sigma_{\gamma\gamma \to \gamma\gamma} = \frac{973}{10125} \frac{\alpha^3}{\pi^2} r_e^2 \left(\frac{\hbar\omega}{m_e c^2} \right)^6, \tag{13}$$

for $\hbar\omega / m_e c^2 < 1$ and reaches its maximum, $\sigma_{\max} \approx 10^{-20}\, cm^2$ for $\hbar\omega \approx m_e c^2$.

Considering formulae (12) and (13) we conclude that linear hyperbolic diffusion equation is valid only for the laser intensities $I \leq 10^{29}$ W/cm². For high intensities the nonlinear hyperbolic diffusion equation must be formulated and solved.

Table 1. Hierarchical structure of the thermal excitation

Interaction	α	mc2α
Electromagnetic	137-1	$0.5/137$
Strong	$\frac{15}{100}$	$\frac{140 \cdot 15}{100}$ for pions
		$\frac{1000 \cdot 15}{100}$ for nucleons
Quark - Quark	1	417*

* D.H. Perkins, Introduction to high energy physics, Addison – Wesley, USA 1987

REFERENCES

[1] Mourou, G. A. et al. *Physics Today* 1998.

[2] Stricland, D.; Mourou, G. A. *Opt. Comm.* 1985, *56*, 219.

Chapter 1

FUNDAMENTALS OF THE LASER PULSES INTERACTION WITH MATTER

1.1. CLASSICAL HYPERBOLIC TRANSPORT EQUATION

As early as 1956 M. Kac considered a particle moving on line at speed c, taking discrete steps of equal size, and undergoing collisions (reversals of direction) at random times, according to a Poisson process of intensity a. He showed that the expected position of the particle satisfies either of two difference equations, according to its initial direction. With correct scaling followed by a passage to the limit, the difference equations become a pair of first order partial differential equations (PDE). Differentiating those and adding them yields the hyperbolic diffusion equation

$$\frac{d^2u}{dt^2}+a\frac{du}{dt}=c\frac{d}{dx}\left(c\frac{du}{dx}\right). \tag{1.1}$$

This is an equation of hyperbolic type. If the lower term (in time) is dropped, it's just the one dimensional wave equation.

R. Hersh proposed the operator generalization of Eq. (1.1):

$$\frac{d}{dt}\left(\frac{dy}{dt}\right)+a\frac{dy}{dt}=A^2u\,. \tag{1.2}$$

In equation (1.2) A is the generator of a group of linear operators acting on a linear space B. Instead of transition moving randomly to the left and right at speed c, the time evolution according to generators A and $*A$ is substituted.

The study and applications of the classical hyperbolic diffusion equation (1.1) covers the thermal processes the stock prices, astrophysics and heavy ion physics.

In this paragraph we will study the ultra-short thermal processes in the framework of the hyperbolic diffusion equation.

When an ultrafast thermal pulse (e. g. femtosecond pulse) interacts with a metal surface, the excited electrons become the main carriers of the thermal energy. For a femtosecond thermal pulse, the duration of the pulse is of the same order as the electron relaxation time. In this case, the hyperbolicity of the thermal energy transfer plays an important role.

Radiation deposition of energy in materials is a fundamental phenomenon to laser processing. It converts radiation energy into material's internal energy, which initiates many thermal phenomena, such as heat pulse propagation, melting and evaporation. The operation of many laser techniques requires an accurate understanding and control of the energy deposition and transport processes. Recently, radiation deposition and the subsequent energy transport in metals have been investigated with picosecond and femtosecond resolutions. Results show that during high-power and short-pulse laser heating, free electrons can be heated to an effective temperature much higher than the lattice temperature, which in turn leads to both a much faster energy propagation process and a much smaller lattice-temperature rise than those predicted from the conventional radiation heating model. Corkum et al. [1.1] found that this electron-lattice nonequilibrium heating mechanism can significantly increase the resistance of molybdenum and copper mirrors to thermal damage during high-power laser irradiation when the laser pulse duration is shorter than one nanosecond. Clemens et al. [1.2] studied thermal transport in multilayer metals during picosecond laser heating. The measured temperature response in the first 20 ps was found to be different from predictions of the conventional Fourier model. Due to the relatively low temporal resolution of the experiment (~ 4 ps), however, it is difficult to determine whether this difference is the result of nonequilibrium laser heating or is due to other heat conduction mechanisms, such as non-Fourier heat conduction, or reflection and refraction of thermal waves at interfaces. Heat is conducted in solids through electrons and phonons. In metals, electrons dominate the heat conduction, while in insulators and semiconductors, phonons are the major heat carriers. Table 1.1 lists important features of the electrons and phonons.

Table 1.1. General Features of Heat Carriers

	Free Electron	Phonon
Generation	ionization or excitation	lattice vibration
Propagation media	vacuum or media	media only
Statistics	Fermion	Boson
Dispersion	$E = \hbar^2 q^2/(2m)$	$E = E(q)$
Velocity ($m \cdot s^{-1}$)	$\sim 10^6$	$\sim 10^3$

The traditional thermal science, or macroscale heat transfer, employs phenomenological laws, such as Fourier's law, without considering the detailed motion of the heat carriers. Decreasing dimensions, however, have brought an increasing need for understanding the heat transfer processes from the microscopic point of view of the heat carriers. The response of the electron and phonon gases to the external perturbation initiated by laser irradiation can be described with the help of a memory function of the system. To that aim, let us consider the generalized Fourier law:

$$q(t) = -\int_{-\infty}^{t} K(t-t')\nabla T(t')dt', \tag{1.3}$$

where $q(t)$ is the density of a thermal energy flux, $T(t')$ is the temperature of electrons and $K(t - t')$ is a memory function for thermal processes. The density of thermal energy flux satisfies the following equation of heat conduction:

$$\frac{\partial}{\partial t}T(t) = \frac{1}{\rho c_v}\nabla^2 \int_{-\infty}^{t} K(t-t')T(t')dt', \tag{1.4}$$

where ρ is the density of charge carriers and c_v is the specific heat of electrons in a constant volume. We introduce the following equation for the memory function describing the Fermi gas of charge carriers:

$$K(t-t') = K_1\delta(t-t'). \tag{1.5}$$

In this case, the electron has a very "short" memory due to thermal disturbances of the state of equilibrium. Combining Eqs. (1.5) and (1.4) we obtain

$$\frac{\partial}{\partial t}T(t) = \frac{1}{\rho c_v}K_1\nabla^2 T. \tag{1.6}$$

Equation (1.6) has the form of the parabolic equation for heat conduction (PHC). Using this analogy, Eq. (1.6) may be transformed as follows:

$$\frac{\partial}{\partial t}T(t) = D_T\nabla^2 T, \tag{1.7}$$

where the heat diffusion coefficient D_T is defined as follows:

$$D_T = \frac{K_1}{\rho c_v}. \tag{1.8}$$

From Eq. (1.8), we obtain the relation between the memory function and the diffusion coefficient

$$K(t-t') = D_T\rho c_v\delta(t-t'). \tag{1.9}$$

In the case when the electron gas shows a "long" memory due to thermal disturbances, one obtains for memory function

$$K(t-t') = K_2 \tag{1.10}$$

When Eq. (1.10) is substituted to the Eq. (1.4) we obtain

$$\frac{\partial}{\partial t}T = \frac{K_2}{\rho c_v}\nabla^2 \int_{-\infty}^{t} T(t')dt'. \tag{1.11}$$

Differentiating both sides of Eq. (1.11) with respect to t, we obtain

$$\frac{\partial^2 T}{\partial t^2} = \frac{K_2}{\rho c_v}\nabla^2 T. \tag{1.12}$$

Equation (1.12) is the hyperbolic wave equation describing thermal wave propagation in a charge carrier gas in a metal film. Using a well-known form of the wave equation,

$$\frac{1}{\upsilon^2}\frac{\partial^2 T}{\partial t^2} = \nabla^2 T. \tag{1.13}$$

and comparing Eqs. (1.12) and (1.13), we obtain the following form for the memory function:

$$K(t-t') = \rho c_v \upsilon^2 \tag{1.14}$$

υ = finite, $\upsilon < \infty$.

As the third case, "intermediate memory" will be considered:

$$K(t-t') = \frac{K_3}{\tau}\exp\left[-\frac{(t-t')}{\tau}\right], \tag{1.15}$$

where τ is the relaxation time of thermal processes. Combining Eqs. (1.15) and (1.4) we obtain

$$c_v\frac{\partial^2 T}{\partial t^2} + \frac{c_v}{\tau}\frac{\partial T}{\partial t} = \frac{K_3}{\rho\tau}\nabla^2 T \tag{1.16}$$

and

$$K_3 = D_\tau c_v \rho. \tag{1.17}$$

Thus, finally,

$$\frac{\partial^2 T}{\partial t^2} + \frac{1}{\tau}\frac{\partial T}{\partial t} = \frac{D_\tau}{\tau}\nabla^2 T. \tag{1.18}$$

Equation (1.18) is the hyperbolic equation for heat conduction (HHC), in which the electron gas is treated as a Fermion gas. The diffusion coefficient D_T can be written in the form:

$$D_T = \frac{1}{3} \upsilon_F^3 \tau, \qquad (1.19)$$

where υ_F is the Fermi velocity for the electron gas in a semiconductor. Applying Eq. (1.19) we can transform the hyperbolic equation for heat conduction, Eq. (1.18), as follows:

$$\frac{\partial^2 T}{\partial t^2} + \frac{1}{\tau} \frac{\partial T}{\partial t} = \frac{1}{3} \upsilon_F^3 \nabla^2 T. \qquad (1.20)$$

Let us denote the velocity of disturbance propagation in the electron gas as s:

$$s = \sqrt{\frac{1}{3}} \upsilon_F. \qquad (1.21)$$

Using the definition of s, Eq. (1.20) may be written in the form

$$\frac{1}{s^2} \frac{\partial^2 T}{\partial t^2} + \frac{1}{\tau s^2} \frac{\partial T}{\partial t} = \nabla^2 T. \qquad (1.22)$$

For the electron gas, treated as the Fermi gas, the velocity of sound propagation is described by the equation

$$\upsilon_s = \left(\frac{P_F^2}{3mm^*} \left(1 + F_0^S\right) \right)^{1/2}, \quad P_F = m\upsilon_F, \qquad (1.23)$$

where m is the mass of a free (non-interacting) electron and m^* is the effective electron mass. Constant F_0^S represents the magnitude of carrier-carrier interaction in the Fermi gas. In the case of a very weak interaction, $m \to m^*$ and $F_0^S \to 0$, so according to Eq. (1.23),

$$\upsilon_S = \frac{m\upsilon_F}{\sqrt{3}m} = \sqrt{\frac{1}{3}} \upsilon_F. \qquad (1.24)$$

To sum up, we can make a statement that for the case of weak electron-electron interaction, sound velocity $\upsilon_S = \sqrt{1/3}\upsilon_F$ and this velocity is equal to the velocity of thermal disturbance propagation s. From this we conclude that the hyperbolic equation for heat conduction Eq. (1.22), is identical as the equation for second sound propagation in the electron gas:

$$\frac{1}{\upsilon_S^2}\frac{\partial^2 T}{\partial t^2}+\frac{1}{\tau\upsilon_S^2}\frac{\partial T}{\partial t}=\nabla^2 T. \tag{1.25}$$

Using the definition expressed by Eq. (1.19) for the heat diffusion coefficient, Eq. (1.25) may be written in the form

$$\frac{1}{\upsilon_S^2}\frac{\partial^2 T}{\partial t^2}+\frac{1}{D_T}\frac{\partial T}{\partial t}=\nabla^2 T. \tag{1.26}$$

The mathematical analysis of Eq. (1.25) leads to the following conclusions:

1. In the case when $\upsilon_S^2 \to \infty$, $\tau\upsilon_S^2$ is finite, Eq. (1.26) transforms into the parabolic equation for heat diffusion:

$$\frac{1}{D_T}\frac{\partial T}{\partial t}=\nabla^2 T. \tag{1.27}$$

2. In the case when $\tau \to \infty$, υ_S is finite, Eq. (1.26) transforms into the wave equation:

$$\frac{1}{\upsilon_S^2}\frac{\partial^2 T}{\partial t^2}=\nabla^2 T. \tag{1.28}$$

Equation (1.28) describes propagation of the thermal wave in the electron gas. From the point of view of theoretical physics, condition $\upsilon_S \to \infty$ violates the special theory of relativity. From this theory we know that there is a limited velocity of interaction propagation and this velocity $\upsilon_{lim} = c$, where c is the velocity of light in a vacuum.

Multiplying both sides of Eq. (1.26) by c^2, we obtain

$$\frac{c^2}{\upsilon_S^2}\frac{\partial^2 T}{\partial t^2}+\frac{c^2}{D_T}\frac{\partial T}{\partial t}=c^2\nabla^2 T, \tag{1.29}$$

Denoting $\beta = \upsilon_S/c$, Eq. (1.29) may be written in the form

$$\frac{1}{\beta^2}\frac{\partial^2 T}{\partial t^2}+\frac{1}{\widetilde{D}_T}\frac{\partial T}{\partial t}=c^2\nabla^2 T, \tag{1.30}$$

where $\widetilde{D}_T=\tau\beta^2$, $\beta<1$. On the basis of the above considerations, we conclude that the heat conduction equation, which satisfies the special theory of relativity, acquires the form of the partial hyperbolic Eq. (1.30). The rejection of the first component in Eq. (1.30) violates the special theory of relativity.

Heat transport during fast laser heating of solids has become a very active research area due to the significant applications of short pulse lasers in the fabrication of sophisticated microstructures, synthesis of advanced materials, and measurements of thin film properties. Laser heating of metals involves the deposition of radiation energy on electrons, the energy exchange between electrons, and the lattice, and the propagation of energy through the media.

The theoretical predictions showed that under ultrafast excitation conditions the electrons in a metal can exist out of equilibrium with the lattice for times of the order of the electron energy relaxation time. Model calculations suggest that it should be possible to heat the electron gas to temperature T_e of up to several thousand degrees for a few picoseconds while keeping the lattice temperature T_l relatively cold. Observing the subsequent equilibration of the electronic system with the lattice allows one to directly study electron-phonon coupling under various conditions.

Several groups have undertaken investigations relating dynamics' changes in the optical constants (reflectivity, transmissivity) to relative changes in electronic temperature. But only recently, the direct measurement of electron temperature has been reported.

The temperature of hot electron gas in a thin gold film (l = 300 Å) was measured, and a reproducible and systematic deviation from a simple Fermi-Dirac (FD) distribution for short time $\Delta t \sim 0.4$ ps were obtained. The nascent electrons are the electrons created by the direct absorption of the photons prior to any scattering.

The relaxation dynamics of the electron temperature with the hyperbolic heat transport equation (HHT), Eq. (1.26), can be investigated. Conventional laser heating processes which involve a relatively low-energy flux and long laser pulse have been successfully modeled in metal processing and in measuring thermal diffusivity of thin films. However, applicability of these models to short-pulse laser heating is questionable. As it is well known, the Anisimov model does not properly take into account the finite time for the nascent electrons to relax to the FD distribution. In the Anisimov model, the Fourier law for heat diffusion in the electron gas is assumed. However, the diffusion equation is valid only when relaxation time is zero, $\tau = 0$, and velocity of the thermalization is infinite, $\upsilon \to \infty$.

The effects of ultrafast heat transport can be observed in the results of front-pump back probe measurements. The results of these types of experiments can be summarized as follows. Firstly, the measured delays are much shorter than would be expected if heat were carried by the diffusion of electrons in equilibrium with the lattice (tens of picoseconds). This suggests that heat is transported via the electron gas alone, and that the electrons are out of equilibrium with the lattice on this time scale. Secondly, since the delay increases approximately linearly with the sample thickness, the heat transport velocity can be extracted, $\upsilon_h \cong 10^8$ cm $\cdot$ s^{-1} = 1μm $\cdot$ ps^{-1}. This is of the same order of magnitude as the Fermi velocity of electrons in gold, 1.4 μm $\cdot$ ps^{-1}.

Since the heat moves at a velocity comparable to υ_F - Fermi velocity of the electron gas, it is natural to question exactly how the transport takes place. Since those electrons which lie close to the Fermi surface are the principal contributors to transport, the heat-carrying electrons move at υ_F. In the limit of lengths longer than the momentum relaxation length, λ, the random walk behavior is averaged and the electron motion is subject to a diffusion equation. Conversely, on a length scale shorter than λ, the electrons move ballistically with velocity close to υ_F. The importance of the ballistic motion may be appreciated by considering the different hot-electron scattering lengths reported in the literature. The electron-electron scattering length in Au, λ_{ee} has been calculated. They find that $\lambda_{ee} \sim (E - E_F$

)^2 for electrons close to the Fermi level. For 2-eV electrons $\lambda_{ee} \approx 35$ nm increase to 80 nm for 1-eV. The electron-phonon scattering length λ_{ep} is usually inferred from conductivity data. Using Drude relaxation times, λ_{ep} can be computed, $\lambda_{ep} \approx 42$ nm at 273 K. This is shorter than λ_{ee}, but of the same order of magnitude. Thus, we would expect that both electron-electron and electron-phonon scattering are important on this length scale. However, since conductivity experiments are steady state measurements, the contribution of phonon scattering in a femtosecond regime experiment such as pump-probe ultrafast lasers, is uncertain. In the usual electron-phonon coupling model, one describes the metal as two coupled subsystems, one for electrons and one for phonons. Each subsystem is in local equilibrium so the electrons are characterized by a FD distribution at temperature T_e and the phonon distribution is characterized by a Bose-Einstein distribution at the lattice temperature T_l. The coupling between the two systems occurs via the electron-phonon interaction. The time evolution of the energies in the two subsystems is given by the coupled parabolic differential equations (Fourier law). For ultrafast lasers, the duration of pump pulse is of the order of relaxation time in metals. In that case, the parabolic heat conduction equation is not valid and hyperbolic heat transport equation must be used (1.26):

$$\frac{1}{\upsilon_S^2}\frac{\partial^2 T}{\partial t^2} + \frac{1}{D_T}\frac{\partial T}{\partial t} = \nabla^2 T, \quad D_T = \tau \upsilon_S^2. \tag{1.31}$$

In Eq. (2.31), υ_S is the thermal wave speed, τ is the relaxation time and D_T denotes the thermal diffusivity. In the following, Eq. (1.31) will be used to describe the heat transfer in the thin gold films.

To that aim, we define: T_e is the electron gas temperature and T_l is the lattice temperature. The governing equations for nonstationary heat transfer are

$$\frac{\partial T_e}{\partial t} = D_T \nabla^2 T - \frac{D_T}{\upsilon_S^2}\frac{\partial^2 T_e}{\partial t^2} - G(T_e - T_l), \qquad \frac{\partial T_l}{\partial t} = G(T_e - T_l). \tag{1.32}$$

where D_T is the thermal diffusivity, T_e is the electron temperature, T_e is the lattice temperature, and G is the electron-phonon coupling constant. In the following, we will assume that on subpicosecond scale the coupling between electron and lattice is weak and Eq. (1.32) can be replaced by the following equations (1.26):

$$\frac{\partial T_e}{\partial t} = D_T \nabla^2 T - \frac{D_T}{\upsilon_S^2}\frac{\partial^2 T_e}{\partial t^2}, \qquad T_l = \text{constant}. \tag{1.33}$$

Equation (1.33) describes nearly ballistic heat transport in a thin gold film irradiated by an ultrafast ($\Delta t < 1$ ps) laser beam. The solution of Eq. (1.33) for 1D is given by [1.3]:

$$T(x,t)=\frac{1}{\upsilon_S}\int dx'T(x',0)\left[\begin{array}{l} e^{-t/2\tau}\dfrac{1}{t_0}\Theta(t-t_0)+ \\ e^{-t/2\tau}\dfrac{1}{2\tau}\left\{\begin{array}{l} I_0\left(\dfrac{(t^2-t_0^2)^{1/2}}{2\tau}\right) \\ +\dfrac{t}{(t^2-t_0^2)^{1/2}}I_1\left(\dfrac{(t^2-t_0^2)^{1/2}}{2\tau}\right)\end{array}\right\}\Theta(t-t_0)\end{array}\right] \tag{1.34}$$

where υ_s is the velocity of second sound, $t_0 = (x - x')/\upsilon_s$ and I_0 and I_1 are modified Bessel functions and $\Theta(t - t_0)$ denotes the Heaviside function. We are concerned with the solution to Eq. (1.34) for a nearly delta function temperature pulse generated by laser irradiation of the metal surface. The pulse transferred to the surface has the shape:

$$\Delta T_0 = \frac{\beta\rho_E}{C_V\upsilon_s\Delta t} \quad \text{for} \quad 0 \le x \le \upsilon_s\Delta t,$$

$$\Delta T_0 = 0 \quad \text{for} \quad x \ge \upsilon_s\Delta t \tag{1.35}$$

In Eq. (1.35), ρ_E denotes the heating pulse fluence, β is the efficiency of the absorption of energy in the solid, $C_V(T_e)$ is electronic heat capacity, and Δt is duration of the pulse. With $t = 0$ temperature profile described by Eq. (1.35) yields:

$$T(l,t)=\frac{1}{2}\Delta T_0 e^{-t/2\tau}\Theta(t-t_0)\Theta(t_0+\Delta t-t) \tag{1.36}$$

$$+\frac{\Delta t}{4\tau}\Delta T_0 e^{-t/2\tau}\left\{I_0(z)+\frac{t}{2\tau}\frac{1}{z}I_1(z)\right\}\Theta(t-t_0),$$

where $z = (t^2 - t_0^2)^{1/2}/2\tau$ and $t = l/\upsilon_s$. The solution to Eq. (1.33), when there are reflecting boundaries, is the superposition of the temperature at l from the original temperature and from image heat source at $\pm 2nl$. This solution is:

$$T(l,t)=\sum_{i=0}^{\infty}\Delta T_0 e^{-t/2\tau}\Theta(t-t_0)\Theta(t_0+\Delta t-t)+\frac{\Delta t}{4\tau}\Delta T_0 e^{-t/2\tau}\left\{I_0(z)+\frac{t}{2\tau}\frac{1}{z}I_1(z)\right\}\Theta(t-t_0), \tag{1.37}$$

where $t_i = t_0, 3t_0, 5t_0, t_0 = l/\upsilon_0$. For gold, $C_V(T_e) = C_e(T_e) = \gamma T_e$, $\gamma = 71.5$ Jm^{-3} K^{-2} and Eq. (1.35) yields:

$$\Delta T_0 = \frac{1.4\times 10^5\rho_E\beta}{\upsilon_s\Delta t T_e} \quad \text{for} \quad 0 \le x \le \upsilon_s\Delta t$$

$$\Delta T_0 = 0 \quad \text{for} \quad x \ge \upsilon_s\Delta t, \tag{1.38}$$

where ρ_E is measured in mJ · cm^{-2}, υ_s in μm · ps^{-1}, and Δt in ps. For $T_e = 300$K :

$$\Delta T_0 = \frac{4.67\times10^2\,\rho_E\beta}{v_s\Delta t} \qquad \text{for} \qquad 0\le x\le \upsilon_s\Delta t$$

$$\Delta T_0 = 0 \qquad \text{for} \qquad x\ge \upsilon_s\Delta t, \tag{1.39}$$

The model calculations (formulae 1.36 – 1.39) were applied to the description of the experimental results and a fairly good agreement of the theoretical calculations and experimental results was obtained.

In the early fifties it was shown by Dingle [1.4], Ward and Wilks [1.5] and London [1.6], that a density fluctuation in a phonon gas would propagate as a thermal wave - a second sound wave - provided that "losses" from the wave were negligible. In one of their papers, Ward and Wilks indicated they would attempt to look for a second sound wave in sapphire crystals. No results of their experiments were published. Then, for nearly a decade, the subject of "thermal wave" lay dormant. Interest was revived in the sixties, primarily through the efforts of J. A. Krumhansl, R. A. Guyer and C. C. Ackerman. The authors measured the thermal wave in dielectric solids was experimentally and theoretically investigated. They found a value for the thermal wave velocity in LiF at a very low temperature $T \sim 1$ K, of $\upsilon_s \sim 100 - 300$ ms^{-1}. In insulators and semiconductors phonons are the major heat carriers. In metals electrons dominate. For long thermal pulses, i.e., when the pulse duration, Δt, is larger than the relaxation time, τ, for thermal processes, $\Delta t >> \tau$, the heat transfer in metals is well described by Fourier diffusion equation. The advent of modern ultrafast lasers opens up the possibility investigating a new mechanism of thermal transport the thermal wave in an electron gas heated by lasers. The effect of an ultrafast heat transport can be observed in the results of front pump back probe measurements. The results of this type of experiments can be summarized as follows. Firstly, the measured delays are much shorter than it would be expected if the heat were carried by the diffusion of electrons in equilibrium with the lattice (tens of picoseconds). This suggests that the heat is transported via the electron gas alone, and that the electrons are out of equilibrium with the lattice within this time scale. Secondly, since the delay increases approximately linearly with the sample thickness, the heat transport velocity can be determined, $\upsilon_h \sim 10^8$ cm s^{-1} = 1μm ps^{-1}. This is of the same order of magnitude as the Fermi velocity of electrons in Au, 1.4 μm ps^{-1}. Kozlowski et al. [1.3] investigated the heat transport in a thin metal film (Au) with the help of the hyperbolic heat conduction equation. It was shown that when the memory of the hot electron gas in metals is taken into account, then the HHT is the dominant equation for heat transfer. The hyperbolic heat conduction equation for heat transfer in an electron gas has the form (1.26)

$$\frac{1}{\left(\frac{1}{3}\upsilon_F^2\right)}\frac{\partial^2 T}{\partial t^2} + \frac{1}{\tau\left(\frac{1}{3}\upsilon_F^2\right)}\frac{\partial T}{\partial t} = \nabla^2 T. \tag{1.40}$$

If we consider an infinite electron gas, then the Fermi velocity can be calculated

$$\upsilon_F \cong bc \tag{1.41}$$

In Eq. (1.41), c is the light velocity in vacuum and $b \sim 10^{-2}$. Considering Eq. (1.41), Eq. (1.40) can be written in a more elegant form:

$$\frac{1}{c^2}\frac{\partial^2 T}{\partial t^2} + \frac{1}{\tau c^2}\frac{\partial T}{\partial t} = \frac{b^2}{3}\nabla^2 T. \tag{1.42}$$

In order to derive the Fourier law from Eq. (1.42), we are forced to break the special theory of relativity and put in Eq. (1.42) $c \to \infty$; $\tau \to 0$. In addition, it can be demonstrated from HHT in a natural way, that in electron gas the heat propagation with velocity $\upsilon_h \sim \upsilon_F$ in the accordance with the results of the pump probe experiments.

Considering the importance of the thermal wave in future engineering applications and simultaneously the lack of the simple physics presentation of the thermal wave for engineering audience in the following we present the main results concerning the wave nature of heat transfer.

Hence, we discuss Eq. (1.42) in more detail. Firstly, we observe that the second derivative term dominates when:

$$c^2(\Delta t)^2 < c^2 \Delta t \tau \tag{1.43}$$

i.e., when $\Delta t < \tau$. This implies that for very short heat pulses we have a hyperbolic wave equation of the form:

$$\frac{1}{c^2}\frac{\partial^2 T}{\partial t^2} = \frac{b^2}{3}\nabla^2 T \tag{1.44}$$

and the velocity of the thermal wave is given by

$$\upsilon_{th} \sim \frac{1}{\sqrt{3}}\frac{c}{b}, \qquad b \sim 10^{-2}. \tag{1.45}$$

The velocity υ_{th} in Eq. (1.45) is the velocity of the thermal wave in an infinite Fermi gas of electrons, which is free of all impurities. The thermal wave, which is described by the solution of Eq. (1.44), does not interact with the crystal lattice. It is the maximum value of the thermal wave obtainable in an infinite free electron gas. If we consider the opposite case to that in Eq. (1.43)

$$c^2(\Delta t)^2 > c^2 \Delta t \tau \tag{1.46}$$

i.e., when

$$\Delta t > \tau \tag{1.47}$$

then, one obtains from Eq. (1.42):

$$\frac{1}{\tau c^2}\frac{\partial T}{\partial t} = \frac{b^2}{3}\nabla^2 T. \tag{1.48}$$

Eq. (1.48) is the parabolic heat conduction equation – Fourier equation.

The value of the thermal wave velocity υ_h is taken from paper [2.20]. Isotherms are presented as a function of the thin film thickness (length) l [μm] and the delay times. The mechanism of heat transfer on a nanometer scale, can be divided into three stages: a heat wave for $t \sim L\upsilon_{th}^{-1}$, mixed heat transport for $L\upsilon_{th}^{-1} < t < 3L\upsilon_{th}^{-1}$ and diffusion for $t > 3L\upsilon_{th}^{-1}$. The thermal wave moves in a manner described by the hyperbolic differential partial equation, $x = \upsilon_{th}\, t$. For $t < x\, \upsilon_{th}^{-1}$ the system is undisturbed by an external heat source (laser beam). For longer heat pulses the evidence of the thermal wave is gradually reduced - but the retardation of the thermal pulse is still evident.

If heat is released in a body of gas liquid or solid, a thermal flux transported by heat conduction appears. The pressure gradients associated with the thermal gradients set a gas or liquid in motion, so that additional energy transport occurs through convection. In particular, at sufficiently large energy releases, shock waves are formed in a gas or liquid which transport thermal energy at velocities larger that the speed of sound. Below the critical energy release, nearly pure thermal wave may propagate owing to heat conduction in a gas or liquid with other transport mechanisms being negligible. Solids metals provide an ideal test medium for the study of thermal waves, since they are practically incompressible at temperature below their melting point and the thermal wave pressures are small compared to the classic pressure (produced by repulsion of the atoms in the lattice) up to large energy releases. In accordance with this picture, the speed of sound in a metal is independent of temperature and given by $c_s = (E/\rho)^{1/2}$ where E is the elasticity modulus and ρ is the density.

Using the path-integral method developed in paper [1.7], the solution of the HHT can be obtained. It occurs that the velocity of the thermal wave in medium is lower than the velocity of the initial thermal wave. The slowing of the thermal wave is caused by the scattering of heat carriers in medium. The scatterings also change the phase of the initial thermal wave.

In one-dimensional flow of heat in metals, the hyperbolic heat transport equation is given by (1.20).

$$\tau\frac{\partial^2 T}{\partial t^2} + \frac{\partial T}{\partial t} = D_T\frac{\partial^2 T}{\partial x^2}, \qquad D_T = \frac{1}{3}\upsilon_F^3\tau, \tag{1.49}$$

where τ denotes the relaxation time, D_T is the diffusion coefficient and T is the temperature. Introducing the non-dimensional spatial coordinate $z = x/ƛ$, where $ƛ = \lambda/2\pi$ denotes the reduced mean free path, Eq. (1.49) can be written in the form:

$$\frac{1}{\upsilon'^2}\frac{\partial^2 T}{\partial t^2} + \frac{2a}{\upsilon'^2}\frac{\partial T}{\partial t} = \frac{\partial^2 T}{\partial z^2}, \tag{1.50}$$

where

$$\upsilon' = \frac{\upsilon}{\lambda} \qquad a = \frac{1}{2\tau}. \tag{1.51}$$

In Eq. (1.51) υ denotes the velocity of heat propagation [1.3], $\upsilon = (D/\tau)^{1/2}$.
Considering the results of the paper [1.7] the solution of Eq. (1.150) obtains the form:
$T(z,0) = \Phi(z)$ an "arbitrary" function

$$\frac{\partial T(z,t)}{\partial t}\Big|_{t=0} = 0 \tag{1.52}$$

the general solution of the Eq. (1.49) has the form:

$$T(z,t) = \frac{1}{2}\left[\Phi(z,t) + \Phi(z,-t)\right]e^{-at}$$
$$+\frac{a}{2}e^{-at}\int_0^t d\eta\left[\Phi(z,\eta) + \Phi(z,-\eta)\right] \tag{1.53}$$
$$+\left[I_0\left(a(t^2-\eta^2)^{1/2}\right) + \frac{t}{(t^2-\eta^2)^{1/2}} I_1\left(a(t^2-\eta^2)^{1/2}\right)\right]$$

In Eq. (1.53), $I_0(x)$ and $I_1(x)$ denote the modified Bessel function of zero and first order respectively.

Let us consider the propagation of the initial thermal wave with velocity υ', i.e.,

$$\Phi(z - \upsilon' t) = \sin(z - \upsilon' t) \tag{1.54}$$

In that case, the integral in (1.53) can be computed analytically, $\Phi(z, t) + \Phi(z, -t) = 2\sin z\cos(\upsilon' t)$ and the integrals on the right-hand side of (1.53) can be done explicitly, we obtain:

$$F(z,t) = e^{-at}\left[\frac{a}{w_1}\sin(w_1 t) + \cos(w_1 t)\right]\sin z, \qquad \upsilon' \geq a \tag{1.55}$$

and

$$F(z,t) = e^{-at}\left[\frac{a}{w_2}\sinh(w_2 t) + \cosh(w_2 t)\right]\sin z, \qquad \upsilon' < a \tag{1.56}$$

where $w_1 = (\upsilon'^2 - a^2)^{1/2}$ and $w_2 = (a^2 - \upsilon'^2)^{1/2}$.

In order to clarify the physical meaning of the solutions given by formulas (1.55) and (1.56), we observe that $\upsilon' = \upsilon/\lambda$ and w_1 and w_2 can be written as:

$$\upsilon_1 = \lambda w_1 = \upsilon\left(1-\left(\frac{1}{2\tau\omega}\right)^2\right)^{1/2}, \qquad 2\tau\omega > 1,$$

$$\upsilon_2 = \lambda w_2 = \upsilon\left(\left(\frac{1}{2\tau\omega}\right)^2 - 1\right)^{1/2}, \qquad 2\tau\omega < 1, \tag{1.57}$$

where ω denotes the pulsation of the initial thermal wave. From formula (1.57), it can be concluded that we can define the new effective thermal wave velocities υ_1 and υ_2. Considering formulas (1.56) and (1.57), we observe that the thermal wave with velocity υ_2 is very quickly attenuated in time. It occurs that when $\omega^{-1} > 2\tau$, the scatterings of the heat carriers diminish the thermal wave.

It is interesting to observe that in the limit of a very short relaxation time, i.e., when $\tau \to 0$, $\upsilon_2 \to \infty$, because for $\tau \to 0$ Eq. (1.49) is the Fourier parabolic equation.

It can be concluded, that for $\omega^{-1} > 2\tau$, the Fourier equation is relevant equation for the description of the thermal phenomena in metals. For $\omega^{-1} > 2\tau$, the scatterings are slower than in the preceding case and attenuation of the thermal wave is weaker. In that case, $\tau \neq 0$ and υ_1 is always finite:

$$\upsilon_1 = \upsilon\left(1-\left(\frac{1}{2\tau\omega}\right)^2\right)^{1/2} < \upsilon. \tag{1.58}$$

For $\tau \to 0$, i.e., for very rare scatterings $\upsilon_1 \to \upsilon$ and Eq. (1.49) is a nearly free thermal wave equation. For τ finite the $\upsilon_1 < \upsilon$ and thermal wave propagates in the medium with smaller velocity than the velocity of the initial thermal wave.

Considering the formula (1.55), one can define the change of the phase of the initial thermal wave β, i.e.:

$$\tan[\beta] = \frac{a}{w_1} = \frac{1}{2\tau\omega}\frac{1}{\sqrt{1-\frac{1}{4\tau^2\omega^2}}}, \qquad 2\tau\omega > 1. \tag{1.59}$$

We conclude that the scatterings produce the change of the phase of the initial thermal wave. For $\tau \to \infty$ (very rare scatterings), $\tan[\beta] = 0$.

According that the complete Schrödinger equation has the form

$$i\hbar\frac{\partial\Psi}{\partial t} = -\frac{\hbar^2}{2m}\nabla^2\Psi + V\Psi, \tag{1.60}$$

where V denotes the potential energy, one can obtain the new parabolic quantum heat transport going back to real time $t \to -2it$ and wave function $\Psi \to T$:

$$\frac{\partial T}{\partial t} = \frac{\hbar}{m}\nabla^2 T - \frac{2V}{\hbar}T. \tag{1.61}$$

Equation (1.61) describer the quantum heat transport for $\Delta t > \tau$, where τ is the relaxation time. For heat transport initiated by ultrashort laser pulses, when $\Delta t > \tau$ one obtains the second order PDE for quantum thermal phenomena

$$\tau\frac{\partial^2 T}{\partial t^2} + \frac{\partial T}{\partial t} = \frac{\hbar}{m}\nabla^2 T - \frac{2V}{\hbar}T. \tag{1.62}$$

Equation (1.62) can be written as

$$\frac{2V\tau}{\hbar}T + \tau\frac{\partial T}{\partial t} + \tau^2\frac{\partial^2 T}{\partial t^2} = \frac{\tau\hbar}{m}\nabla^2 T. \tag{1.63}$$

For distortionless thermal phenomena we obtain

$$V\tau \approx \frac{\hbar}{2}. \tag{1.64}$$

Equation (1.64) is Heisenberg uncertainty relation for thermal quantum phenomena. Substituting equation (1.64) to equation (1.63) we obtain the new form of quantum thermal equation

$$\left(1 + \tau\frac{\partial}{\partial t} + \tau^2\frac{\partial^2}{\partial t^2}\right)T = \frac{\tau\hbar}{m}\nabla^2 T. \tag{1.65}$$

It is obvious, from a dimensional analysis, that one can add the fourth term in equation (1.65), i.e.

$$\left(1 + \tau\frac{\partial}{\partial t} + \tau^2\frac{\partial^2}{\partial t^2} + \tau^3\frac{\partial^3}{\partial t^3}\right)T = \frac{\tau\hbar}{m}\nabla^2 T. \tag{1.66}$$

when $V = 0$ equation (1.66) has the form

$$\left(\tau\frac{\partial}{\partial t} + \tau^2\frac{\partial^2}{\partial t^2} + \tau^3\frac{\partial^3}{\partial t^3}\right)T = \frac{\tau\hbar}{m}\nabla^2 T. \tag{1.67}$$

Let us write Eq. (1.67) in the form

$$\kappa \nabla^2 T = \varepsilon \frac{\partial^2 T}{\partial t^2} + \mu \frac{\partial T}{\partial t} + \mu_3 \frac{\partial^3 T}{\partial t^3}. \tag{1.68}$$

where

$$\kappa = \frac{\tau \hbar}{m}, \qquad \varepsilon = \tau^2, \qquad \mu = \tau, \qquad \mu_3 = \tau^3 . \tag{1.69}$$

Equation (1.68) yields the characteristic polynomial equation

$$p(s, jk) = \mu_3 s^3 + \varepsilon s^2 + \mu s + \kappa k^2 = 0. \tag{1.70}$$

Equation (1.68) was investigated, for oscillating transport phenomena, by P.M. Ruiz. In one-dimensional case one obtains from Eq. (1.67)

$$\tau \frac{\partial T}{\partial t} + \tau^2 \frac{\partial^2 T}{\partial t^2} + \tau^3 \frac{\partial^3 T}{\partial t^3} = \frac{\tau \hbar}{m} \frac{\partial^2 T}{\partial x^2}. \tag{1.71}$$

Below we analyze the third-order wave equation

$$\tau^2 \frac{\partial^3 T}{\partial t^3} = \frac{\hbar}{m} \frac{\partial^2 T}{\partial x^2}. \tag{1.72}$$

In the case of thermal processes induced by attosecond laser pulses

$$\tau = \frac{\hbar}{m\upsilon^2}, \qquad \upsilon = \alpha c, \tag{1.73}$$

and equation (1.72) can be rewritten as

$$\frac{\partial^2 T}{\partial x^2} = \beta \frac{\partial^3 T}{\partial t^3}, \qquad \beta = \frac{\hbar}{m\upsilon^4}. \tag{1.74}$$

We seek a solution of equation (1.74) of the form

$$T(x,t) = Ae^{i(kx - \omega t)}. \tag{1.75}$$

Substituting equation (1.75) to Eq. (1.74) one obtains

$$(ik)^2 = \beta(-i\omega)^3. \tag{1.76}$$

This shows that equation (1.75) is the solution of the third-order PDE (1.74) i.e. Eq. (1.74) is the third-order wave equation if

$$\beta = \frac{(ik)^2}{(-i\omega)^3} = \frac{\hbar}{m\upsilon^4}, \tag{1.77}$$

where υ is the speed of propagation of thermal energy (1.33). Substituting Eq. (1.77) to Eq. (1.75) one obtains

$$T(x,t) = Ae^{i\left[\frac{x}{\sqrt{2}\lambda}-\omega t\right]} e^{-\frac{1}{\sqrt{2}}\frac{x}{\lambda}} + Be^{-i\left[\frac{x}{\sqrt{2}\lambda}-\omega t\right]} e^{\frac{1}{\sqrt{2}}\frac{x}{\lambda}}, \tag{1.78}$$

where λ is mean free path. The second term in Eq. (1.78) tends to infinity for $\frac{x}{\lambda} >> 1$ and is to be omitted. The final solution of Eq. (1.75) has the form

$$T(x,t) = e^{-\frac{1}{\sqrt{2}}\frac{x}{\lambda}} Ae^{i\left[\frac{x}{\sqrt{2}\lambda}-\omega t\right]} \tag{1.79}$$

and describes the strongly damped thermal wave.

It is interesting to observe that for electromagnetic interaction the third-order time derivative $\frac{\partial^3 x}{\partial t^3}$ also describes the damping of the electron motion due to the self interaction of the charges.

1.2. Quantum Hyperbolic Transport Equation

Dynamical processes are commonly investigated using laser pump-probe experiments with a pump pulse exciting the system of interest and a second probe pulse tracking is temporal evolution. As the time resolution attainable in such experiments depends on the temporal definition of the laser pulse, pulse compression to the attosecond domain is a recent promising development.

After the standards of time and space were defined the laws of classical physics relating such parameters as distance, time, velocity, temperature are assumed to be independent of accuracy with which these parameters can be measured. It should be noted that this assumption does not enter explicitly into the formulation of classical physics. It implies that together with the assumption of existence of an object and really independently of any measurements (in classical physics) it was tacitly assumed that *there was a possibility of an unlimited increase in accuracy of measurements.* Bearing in mind the "atomicity" of time i.e. considering the smallest time period, the Planck time, the above statement is obviously not true. Attosecond laser pulses we are at the limit of laser time resolution.

With attosecond laser pulses belong to a new Nano – World where size becomes comparable to atomic dimensions, where transport phenomena follow different laws from that in the macro world. This first stage of miniaturization, from 10^{-3} m to 10^{-6} m is over and the new one, from 10^{-6} m to 10^{-9} m just beginning. The Nano – World is a quantum world with all the predicable and non-predicable (yet) features.

In this paragraph, we develop and solve the quantum relativistic heat transport equation for Nano – World transport phenomena where external forces exist.

There is an impressive amount of literature on hyperbolic heat transport in matter. In Chapter 1.1 we developed the new hyperbolic heat transport equation which generalizes the Fourier heat transport equation for the rapid thermal processes. The hyperbolic heat transport equation (HHT) for the fermionic system has be written in the form (1.25)

$$\frac{1}{\left(\frac{1}{3}\upsilon_F^2\right)}\frac{\partial^2 T}{\partial t^2}+\frac{1}{\tau\left(\frac{1}{3}\upsilon_F^2\right)}\frac{\partial T}{\partial t}=\nabla^2 T\ , \tag{1.80}$$

where T denotes the temperature, τ the relaxation time for the thermal disturbance of the fermionic system, and υ_F is the Fermi velocity.

In what follows we develop the new formulation of the HHT, considering the details of the two fermionic systems: electron gas in metals and the nucleon gas.

For the electron gas in metals, the Fermi energy has the form [1.3]

$$E_F^e=(3\pi)^2\frac{n^{2/3}\hbar^2}{2m_e}, \tag{1.81}$$

where n denotes the density and m_e electron mass. Considering that

$$n^{-1/3}\sim a_B\sim\frac{\hbar^2}{me^2}, \tag{1.82}$$

and a_B = Bohr radius, one obtains

$$E_F^e\sim\frac{n^{2/3}\hbar^2}{2m_e}\sim\frac{\hbar^2}{ma^2}\sim\alpha^2 m_e c^2, \tag{1.83}$$

where c = light velocity and α = 1/137 is the fine-structure constant for electromagnetic interaction. For the Fermi momentum p_F we have

$$p_F^e\sim\frac{\hbar}{a_B}\sim\alpha m_e c, \tag{1.84}$$

and, for Fermi velocity υ_F ,

$$v_F^e \sim \frac{p_F}{m_e} \sim \alpha c. \tag{1.85}$$

Formula (1.85) gives the theoretical background for the result presented in Chapter 1.1. Comparing formulas (1.41) and (1.85) it occurs that $b = \alpha$. Considering formula (1.85), Eq. (1.42) can be written as

$$\frac{1}{c^2}\frac{\partial^2 T}{\partial t^2} + \frac{1}{c^2 \tau}\frac{\partial T}{\partial t} = \frac{\alpha^2}{3}\nabla^2 T. \tag{1.86}$$

As seen from (1.86), the HHT equation is a relativistic equation, since it takes into account the finite velocity of light.

For the nucleon gas, Fermi energy equals

$$E_F^N = \frac{(9\pi)^{2/3}\hbar^2}{8mr_0^2}, \tag{1.87}$$

where m denotes the nucleon mass and r_0, which describes the range of strong interaction, is given by

$$r_0 = \frac{\hbar}{m_\pi c}, \tag{1.88}$$

wherein m_π is the pion mass. From formula (1.88), one obtains for the nucleon Fermi energy

$$E_F^N \sim \left(\frac{m_\pi}{m}\right)^2 mc^2. \tag{1.89}$$

In analogy to the Eq. (1.83), formula (1.89) can be written as

$$E_F^N \sim \alpha_s^2 mc^2, \tag{1.90}$$

where $\alpha_s = \frac{m_\pi}{m} \cong 0.15$ is the fine-structure constant for strong interactions. Analogously, we obtain the nucleon Fermi momentum

$$p_F^e \sim \frac{\hbar}{r_0} \sim \alpha_s mc \tag{1.91}$$

and the nucleon Fermi velocity

$$v_F^N \sim \frac{pF}{m} \sim \alpha_s c, \tag{1.92}$$

and HHT for nucleon gas can be written as

$$\frac{1}{c^2}\frac{\partial^2 T}{\partial t^2}+\frac{1}{c^2\tau}\frac{\partial T}{\partial t}=\frac{\alpha_s^2}{3}\nabla^2 T. \tag{1.93}$$

In the following, the procedure for the discretization of temperature $T(\vec{r},t)$ in hot fermion gas will be developed. First of all, we introduce the reduced de Broglie wavelength

$$\begin{aligned} \lambda_B^e &= \frac{\hbar}{m_e v_h^e}, & v_h^e &= \frac{1}{\sqrt{3}}\alpha c, \\ \lambda_B^N &= \frac{\hbar}{m v_h^N}, & v_h^N &= \frac{1}{\sqrt{3}}\alpha_s c, \end{aligned} \tag{1.94}$$

and the mean free paths λ_e and λ_N

$$\lambda^e = v_h^e \tau^e, \quad \lambda^N = v_h^N \tau^N. \tag{1.95}$$

In view of formulas (1.94) and (1.95), we obtain the HHC for electron and nucleon gases

$$\frac{\lambda_B^e}{v_h^e}\frac{\partial^2 T}{\partial t^2}+\frac{\lambda_B^e}{\lambda^e}\frac{\partial T}{\partial t}=\frac{\hbar}{m_e}\nabla^2 T^e, \tag{1.96}$$

$$\frac{\lambda_B^N}{v_h^N}\frac{\partial^2 T}{\partial t^2}+\frac{\lambda_B^N}{\lambda^N}\frac{\partial T}{\partial t}=\frac{\hbar}{m}\nabla^2 T^N. \tag{1.97}$$

Equations (1.96) and (1.97) are the hyperbolic partial differential equations which are the master equations for heat propagation in Fermi electron and nucleon gases. In the following, we will study the quantum limit of heat transport in the fermionic systems. We define the quantum heat transport limit as follows:

$$\lambda^e = \lambda_B^e, \quad \lambda^N = \lambda_B^N. \tag{1.98}$$

In that case, Eqs. (1.96) and (1.97) have the form

$$\tau^e\frac{\partial^2 T^e}{\partial t^2}+\frac{\partial T^e}{\partial t}=\frac{\hbar}{m_e}\nabla^2 T^e, \tag{1.99}$$

$$\tau^N\frac{\partial^2 T^N}{\partial t^2}+\frac{\partial T^N}{\partial t}=\frac{\hbar}{m}\nabla^2 T^N, \tag{1.100}$$

where

$$\tau^e = \frac{\hbar}{m_e\left(v_h^e\right)^2}, \qquad \tau^N = \frac{\hbar}{m\left(v_h^N\right)^2}. \tag{1.101}$$

Equations (1.99) and (1.100) define the master equation for quantum heat transport (QHT). Having the relaxation times τ^e and τ^N, one can define the "pulsations" ω_h^e and ω_h^N

$$\omega_h^e = (\tau^e)^{-1}, \qquad \omega_h^N = (\tau^N)^{-1}, \tag{1.102}$$

or

$$\omega_h^e = \frac{m_e\left(v_h^e\right)^2}{\hbar}, \qquad \omega_h^N = \frac{m\left(v_h^N\right)^2}{\hbar},$$

i.e.,

$$\begin{aligned} \omega_h^e \hbar &= m_e\left(v_h^e\right)^2 = \frac{m_e\alpha^2}{3}c^2, \\ \omega_h^N \hbar &= m\left(v_h^N\right)^2 = \frac{m\alpha_s^2}{3}c^2. \end{aligned} \tag{1.103}$$

The formulas (1.103) define the Planck-Einstein relation for heat quanta E_h^e and E_h^N

$$\begin{aligned} E_h^e &= \omega_h^e \hbar = m_e\left(v_h^e\right)^2, \\ E_h^N &= \omega_h^N \hbar = m_N\left(v_h^N\right)^2. \end{aligned} \tag{1.104}$$

The heat quantum with energy $E_h = \hbar\omega$ can be named the *heaton*, in complete analogy to the *phonon*, *magnon*, *roton*, etc. For τ^e, $\tau^N \to 0$, Eqs. (1.99) and (1.103) are the Fourier equations with quantum diffusion coefficients D^e and D^N

$$\frac{\partial T^e}{\partial t} = D^e \nabla^2 T^e, \qquad D^e = \frac{\hbar}{m_e}, \tag{1.105}$$

$$\frac{\partial T^N}{\partial t} = D^N \nabla^2 T^N, \qquad D^N = \frac{\hbar}{m}. \tag{1.106}$$

The quantum diffusion coefficients D^e and D^N were introduced for the first time by E. Nelson.

For finite τ^e and τ^N, for $\Delta t < \tau^e$, $\Delta t < \tau^N$, Eqs. (1.99) and (1.100) can be written as

$$\frac{1}{(v_h^e)^2}\frac{\partial^2 T^e}{\partial t^2} = \nabla^2 T^e, \tag{1.107}$$

$$\frac{1}{(v_h^N)^2}\frac{\partial^2 T^N}{\partial t^2} = \nabla^2 T^N. \tag{1.108}$$

Equations (1.107) and (1.108) are the wave equations for quantum heat transport (QHT). For $\Delta t > \tau$, one obtains the Fourier equations (1.105) and (1.106).

In what follows, the dimensionless form of the QHT will be used. Introducing the reduced time t' and reduced length x',

$$t' = t/\tau, \qquad x' = \frac{x}{v_h \tau}, \tag{1.109}$$

one obtains, for QHT,

$$\frac{\partial^2 T^e}{\partial t^2} + \frac{\partial T^e}{\partial t} = \nabla^2 T^e, \tag{1.110}$$

$$\frac{\partial^2 T^N}{\partial t^2} + \frac{\partial T^N}{\partial t} = \nabla^2 T^N \tag{1.111}$$

and, for QFT,

$$\frac{\partial T^e}{\partial t} = \nabla^2 T^e, \tag{1.112}$$

$$\frac{\partial T^N}{\partial t} = \nabla^2 T^N. \tag{1.113}$$

1.3. Heat Transport in Nanoscale

Clusters and aggregates of atoms in the nanometer range (currently called nanoparticles) are systems intermediate in several respects, between simple molecules and bulk materials and have been the subject of intensive work.

In this paragraph, we investigate the thermal relaxation phenomena in nanoparticles – microtubules within the frame of the quantum heat transport equation. In reference [1.3], the thermal inertia of materials, heated with laser pulses faster than the characteristic relaxation time was investigated. It was shown, that in the case of the ultra-short laser pulses it was necessary to use the hyperbolic heat conduction (HHC). For microtubules the diameters are of the order of the de Broglie wave length. In that case quantum heat transport must be usedto describe the transport phenomena,

$$\tau\frac{\partial^2 T}{\partial t^2}+\frac{\partial T}{\partial t}=\frac{\hbar}{m}\nabla^2 T, \tag{1.114}$$

where T denotes the temperature of the heat carrier, and m denotes its mass and τ is the relaxation time. The relaxation time τ is defined as:

$$\tau=\frac{\hbar}{m\upsilon_h^2}, \tag{1.115}$$

where υ_h is the thermal pulse propagation rate

$$\upsilon_h=\frac{1}{\sqrt{3}}\alpha c. \tag{1.116}$$

In equation (1.116) α is a coupling constant (for the electromagnetic interaction $\alpha = e^2/\hbar c$ and c denotes the speed of light in vacuum. Both parameters τ and υ_h characterizes completely the thermal energy transport on the atomic scale and can be termed *"atomic relaxation time"* and *"atomic"* heat diffusivity.

Both τ and υ_h contain constants of Nature, α, c. Moreover, on an atomic scale there is no shorter time period than and smaller velocity than that build from of constants in Nature. Consequently, one can call τ and υ_h the *elementary relaxation time* and *elementary diffusivity*, which characterizes heat transport in the elementary building block of matter, the atom. In the following, starting with elementary τ and υ_h, we shall describe thermal relaxation processes in microtubules which consist of the N components (molecules) each with elementary τ and υ_h. With this in view, we use the Pauli-Heisenberg inequality [1.3, 1.8]

$$\Delta r\Delta p \geq N^{\frac{1}{3}}\hbar, \tag{1.117}$$

where r denotes the characteristic dimension of the nanoparticle and p is the momentum of the energy carriers. The Pauli-Heisenberg inequality expresses the basic property of the N – fermionic system. In fact, compared to the standard Heisenberg inequality

$$\Delta r\Delta p \geq \hbar, \tag{1.118}$$

we observe that, in this case the presence of the large number of identical fermions forces the system either to become spatially more extended for a fixed typical momentum dispension, or to increase its typical momentum dispension for a fixed typical spatial extension. We could also say that for a fermionic system in its ground state, the average energy per particle increases with the density of the system.

An illustrative means of interpreting the Pauli-Heisenberg inequality is to compare Eq. (1.117) with Eq. (1.118) and to think of the quantity on the right hand side of it as the *effective fermionic Planck constant*

$$\hbar^{f}(N) = N^{\frac{1}{3}}\hbar. \tag{1.119}$$

We could also say that antisymmetrization, which typifies fermionic amplitudes amplifies those quantum effects which are affected by the Heisenberg inequality.

Based on equation (1.119), we can recalculate the relaxation time τ, equation (1.115) and the thermal speed υ_h, equation (1.116) for a nanoparticle consisting of N fermions

$$\hbar \leftarrow \hbar^{f}(N) = N^{\frac{1}{3}}\hbar \tag{1.120}$$

and obtain

$$\upsilon_h^f = \frac{e^2}{\hbar^f(N)} = \frac{1}{N^{\frac{1}{3}}}\upsilon_h, \tag{1.121}$$

$$\tau^f = \frac{\hbar^f}{m\left(\upsilon_h^f\right)^2} = N\tau. \tag{1.122}$$

The number N particles in a nanoparticle (sphere with radius r) can be calculated using the equation (we assume that density of a nanoparticle does not differ too much from that of the bulk material)

$$N = \frac{\frac{4\pi}{3}r^3\rho AZ}{\mu} \tag{1.123}$$

and for non spherical shapes with semi axes a, b, c

$$N = \frac{\frac{4\pi}{3}abc\rho AZ}{\mu}, \tag{1.124}$$

where ρ is the density of the nanoparticle, A is the Avogardo number, μ is the molecular mass of theparticles in grams and Z is the number of valence electrons.

Using equations (1.121) and (1.122), we can calculate the de Broglie wave length λ_B^f and mean free path λ_{mfp}^f for nanoparticles

$$\lambda_B^f = \frac{\hbar^f}{m\upsilon_{th}^f} = N^{\frac{2}{3}}\lambda_B,$$

(1.125)

$$\lambda_{emfp}^f = \upsilon_{th}^f \tau_{th}^f = N^{\frac{2}{3}}\lambda_{mfp},$$

(1.126)

where λ_B and λ_{mfp} denote the de Broglie wave length and the mean free path for heat carriers in nanoparticles (e.g. microtubules). Microtubules are essential to cell functions. In neurons, microtubules help and regulate synaptic activity responsible for learning and cognitive function. Whereas microtubules have traditionally been considered to be purely structural elements, recent evidence has revealed that mechanical, chemical and electrical signaling and a communication function also exist as a result of the microtubule interaction with membrane structures by linking proteins, ions and voltage fields respectively. The characteristic dimensions of the microtubules; a crystalline cylinder 10 nm internal diametre, are of the order of the de Broglie length for electrons in atoms. When the characteristic length of the structure is of the order of the de Broglie wave length, then the signaling phenomena must be described by the quantum transport theory. In order to describe quantum transport phenomena in microtubules it is necessary to use equation (1.114) with the relaxation time described by equation

$$\tau = \frac{2\hbar}{m\upsilon^2} = \frac{\hbar}{E}.$$

(1.127)

The relaxation time is the de-coherence time, i.e. the time before the wave function collapses, when the transition classical $\rightarrow$ quantum phenomena is considered.

In the following we consider the time τ for atomic and multiatomic phenomena.

$$\tau_a \approx 10^{-17}\,\mathrm{s}$$

(1.128)

and when we consider multiatomic transport phenomena, with N equal number of aggregates involved the equation is (1.122)

$$\tau_N = N\tau_a.$$

(1.129)

The Penrose – Hameroff Orchestrated Objective Reduction Model (OrchOR) proposes that quantum superposition – computation occurs in nanotubule automata within brain neurons and glia. Tubulin subunits within microtubules act as qubits, switching between states on a nanosecond scale, governed by London forces in hydrophobic pockets. These oscillations are tuned and orchestrated by microtubule associated proteins (MAPs) providing a feedback loop between the biological system and the quantum state. These qubits interact computationally by non-localquantum entanglement, according to the Schrödinger equation with preconscious processing continuing until the threshold for objective reduction (OR) is reached $(E = \hbar/T)$. At that instant, collapse occurs, triggering a "moment of awareness" or a conscious event – an event that determines particular configurations of Planck scale experiential geometry and corresponding classical states of nanotubules automata that regulate synaptic and other neural functions. A sequence of such events could provide a forward flow of subjective time and stream of consciousness. Quantum states in nanotubules may link to those in nanotubules in other neurons and glia by tunneling through gap functions,

Table 1.2. The de-coherence relaxation time

Event	T [ms]	E	N - number of aggregates	T [ms]
Buddhist moment of awareness nucleons	13	4·1015	1015	10
Coherent 40 Hz oscillations	25	2·1015	1015	10
EEG alpha rhytm (8 to 12 Hz)	100	1014	1014	1
Libet's sensory threshold	100	1014	1014	1

permitting extension of the quantum state through significant volumes of the brain.

Based on $E = \hbar / T$, the size and extension of Orch OR events which correlate with a subjective or neurophysiological description of conscious events can be calculated. In Table 1.2 the calculated T (Penrose-Hameroff) and τ –equation (1.129) are presented.

We shall now develop the generalized quantum heat transport equation for microtubules which also includes the potential term. Thus, we are able to use the analogy of the Schrödinger and quantum heat transport equations. If we consider, for the moment, the parabolic heat transport equation with the second derivative term omitted

$$\frac{\partial T}{\partial t} = \frac{\hbar}{m} \nabla^2 T. \tag{1.130}$$

If the real time $t \to it/2$, $T \to \Psi$, Eq. (1.130) has the form of a free Schrödinger equation

$$i\hbar \frac{\partial \Psi}{\partial t} = -\frac{\hbar^2}{2m} \nabla^2 \Psi. \tag{1.131}$$

The complete Schrödinger equation has the form

$$i\hbar \frac{\partial \Psi}{\partial t} = -\frac{\hbar^2}{2m} \nabla^2 \Psi + V\Psi, \tag{1.132}$$

where V denotes the potential energy. When we go back to real time $t \to 2it$, $\Psi \to T$, the new parabolic heat transport is obtained

$$\frac{\partial T}{\partial t} = \frac{\hbar}{m}\nabla^2 T - \frac{2V}{\hbar}T. \tag{1.133}$$

Equation (1.133) describes the quantum heat transport for $\Delta t > \tau$. For heat transport initiated by ultra-short laser pulses, when $\Delta t < \tau$ one obtains the generalized quantum hyperbolic heat transport equation

$$\tau\frac{\partial^2 T}{\partial t^2} + \frac{\partial T}{\partial t} = \frac{\hbar}{m}\nabla^2 T - \frac{2V}{\hbar}T. \tag{1.134}$$

Considering that $\tau = \hbar/m\upsilon^2$, Eq. (1.134) can be written as follows:

$$\frac{1}{\upsilon^2}\frac{\partial^2 T}{\partial t^2} + \frac{m}{\hbar}\frac{\partial T}{\partial t} + \frac{2Vm}{\hbar^2}T = \nabla^2 T. \tag{1.135}$$

Equation (1.135) describes the heat flow when apart from the temperature gradient, the potential energy V (is present.)

In the following, we consider one-dimensional heat transfer phenomena, i.e

$$\frac{1}{\upsilon^2}\frac{\partial^2 T}{\partial t^2} + \frac{m}{\hbar}\frac{\partial T}{\partial t} + \frac{2Vm}{\hbar^2}T = \frac{\partial^2 T}{\partial x^2}. \tag{1.136}$$

We seek a solution in the form

$$T(x,t) = e^{-t/2\tau}u(x,t) \tag{1.137}$$

for the quantum heat transport equation (1.136)

After substitution of Eq. (1.137) into Eq. (1.136), one obtains

$$\frac{1}{\upsilon^2}\frac{\partial^2 u}{\partial t^2} - \frac{\partial^2 u}{\partial x^2} + q^2 u(x,t) = 0, \tag{1.138}$$

where

$$q^2 = \frac{2Vm}{\hbar^2} - \left(\frac{m\upsilon}{2\hbar}\right)^2. \tag{1.139}$$

In the following, we consider a constant potential energy $V = V_0$. The general solution of Eq. (1.138) for the Cauchy boundary conditions,

$$u(x,0) = f(x), \qquad \left[\frac{\partial u(x,t)}{\partial t}\right]_{t=0} = F(x), \tag{1.140}$$

has the form [1.3]

$$u(x,t) = \frac{f(x-\upsilon t) + f(x+\upsilon t)}{2} + \frac{1}{2\upsilon}\int_{x-\upsilon t}^{x+\upsilon t}\Phi(x,y,z)dz, \tag{1.141}$$

where

$$\Phi(x,t,z) = \frac{1}{\upsilon}J_0\left(\frac{b}{\upsilon}\sqrt{(z-x)^2-\upsilon^2t^2}\right) + btf(z)\frac{J_0\left(\frac{b}{\upsilon}\sqrt{(z-x)^2-\upsilon^2t^2}\right)}{\sqrt{(z-x)^2-\upsilon^2t^2}}, \tag{1.142}$$

$$b = \left(\frac{m\upsilon^2}{2\hbar}\right)^2 - \frac{2Vm}{\hbar^2}\upsilon^2$$

and $J_0(z)$ denotes the Bessel function of the first kind. Considering equations (1.137), (1.138), (1.139) the solution of Eq. (1.136) describes the propagation of the distorted thermal quantum waves with characteristic lines $x = \pm \upsilon t$. We can define the distortionless thermal wave as the wave which preserves the shape in the potential energy V_0 field. The condition for conserving the shape can be expressed as

$$q^2 = \frac{2Vm}{\hbar^2} - \left(\frac{m\upsilon}{2\hbar}\right)^2. \tag{1.143}$$

When Eq. (1.143) holds, Eq. (1.138) has the form

$$\frac{\partial^2 u(x,t)}{\partial t^2} = \upsilon^2\frac{\partial^2 u}{\partial x^2}. \tag{1.144}$$

Equation (1.138) is the quantum wave equation with the solution (for Cauchy boundary conditions (1.140))

$$u(x,t) = \frac{f(x-\upsilon t) + f(x+\upsilon t)}{2} + \frac{1}{2\upsilon}\int_{x-\upsilon t}^{x+\upsilon t}F(z)dz. \tag{1.145}$$

It is interesting to observe, that condition (1.143) has an analog in classical theory of the electrical transmission line. In the context of the transmission of an electromagnetic field, the condition $q^2 = 0$ describes the Heaviside distortionless line. Eq. (1.143) – the distortionless condition – can be written as

$$V_0\tau \approx \hbar. \tag{1.146}$$

We can conclude, that in the presence of the potential energy V_0 one can observe the undisturbed quantum thermal wave in microtubules only when *the Heisenberg uncertainty relation for thermal processes* (1.146) is fulfilled.

The generalized quantum heat transport equation (GQHT) (1.136) leads to generalized Schrödinger equation for microtubules. After the substitution $t \to it/2,\ T \to \Psi$ in Eq. (1.136), one obtains the generalized Schrödinger equation (GSE)

$$i\hbar\frac{\partial\Psi}{\partial t} = -\frac{\hbar^2}{2m}\nabla^2\Psi + V\Psi - 2\tau\hbar\frac{\partial^2\Psi}{\partial t^2}. \tag{1.147}$$

Considering that $\tau = \hbar / mv^2 = \hbar / m\alpha^2 c^2$ (α = 1/137) is the fine-structure constant for electromagnetic interactions) Eq. (1.147) can be written as

$$i\hbar\frac{\partial\Psi}{\partial t} = -\frac{\hbar^2}{2m}\nabla^2\Psi + V\Psi - \frac{2\hbar^2}{m\alpha^2c^2}\frac{\partial^2\Psi}{\partial t^2}. \tag{1.148}$$

One can conclude, that for a time period $\Delta t < \hbar / m\alpha^2 c^2 \approx 10^{-17}$ s the description of quantum phenomena needs some revision. On the other hand, for $\Delta t > 10^{-17}$ in GSE the second derivative term can be omitted and as a result the Schrödinger equation SE is obtained, i.e.

$$i\hbar\frac{\partial\Psi}{\partial t} = -\frac{\hbar^2}{2m}\nabla^2\Psi + V\Psi. \tag{1.149}$$

It is interesting to observe, that GSE was discussed also in the context of the sub-quantal phenomena.

In conclusion a study of the interactions of the attosecond laser pulses with matter can shed light on the applicability of the SE in a study of ultra-short sub-quantal phenomena.

The structure of Eq. (1.138) depends on the sign of the parameter q^2. For quantum heat transport phenomena with electrons as the heat carriers the parameter q^2 is a function of the potential barrier height V_0 and velocity v.

The initial Cauchy condition

$$u(x,0) = f(x), \qquad \frac{\partial u(x,0)}{\partial t} = g(x), \tag{1.150}$$

and the solution of the Eq. (1.138) has the form [1.8] and Appendix A:

$$u(x,t)=\frac{f(x-\upsilon t)+f(x+\upsilon t)}{2}+\frac{1}{2\upsilon}\int_{x-\upsilon t}^{x+\upsilon t}g(\varsigma)I_0\left[\sqrt{-q^2(\upsilon^2t^2-(x-\varsigma)^2)}\right]d\varsigma-\frac{\upsilon\sqrt{-q^2}t}{2}\int_{x-\upsilon t}^{x+\upsilon t}f(\varsigma)\frac{I_1\left[\sqrt{-q^2(\upsilon^2t^2-(x-\varsigma)^2)}\right]}{\sqrt{\upsilon^2t^2-(x-\varsigma)^2}}d\varsigma. \quad (1.151)$$

When $q^2 > 0$ Eq. (1.138) is the *Klein – Gordon equation* (K-G), which is well known from applications in elementary particle and nuclear physics.

For the initial Cauchy condition (1.150), the solution of the (K-G) equation can be written as [1.8]

$$u(x,t)=\frac{f(x-\upsilon t)+f(x+\upsilon t)}{2}+\frac{1}{2\upsilon}\int_{x-\upsilon t}^{x+\upsilon t}g(\varsigma)J_0\left[\sqrt{q^2(\upsilon^2t^2-(x-\varsigma)^2)}\right]d\varsigma-\frac{\upsilon\sqrt{q^2}t}{2}\int_{x-\upsilon t}^{x+\upsilon t}f(\varsigma)\frac{J_0'\left[\sqrt{q^2(\upsilon^2t^2-(x-\varsigma)^2)}\right]}{\sqrt{\upsilon^2t^2-(x-\varsigma)^2}}d\varsigma. \quad (1.152)$$

Both solutions (1.151) and (1.152) exhibit the domains of dependence and influence on the *modified Klein-Gordon* and *Klein-Gordon equation.* These domains, which characterize the maximum speed at which a thermal disturbance can travel are determined by the principal terms of the given equation (i.e., the second derivative terms) and do not depend on the lower order terms. It can be concluded that these equations and the wave equation (for m = 0) have identical domains of dependence and influence.

Vacuum energy is a consequence of the quantum nature of the electromagnetic field, which is composed of photons. A photon of frequency ω has an energy $\hbar\omega$, where $\hbar$ is Planck constant. The quantum vacuum can be interpreted as the lowest energy state (or ground state) of the electromagnetic (EM) field which result when all charges and currents have been removed and the temperature has been reduced to absolute zero. In this state no ordinary photons are present. Nevertheless, because the electromagnetic field is a quantum system the energy of the ground state of the EM is not zero. Although the average value of the electric field $\langle E\rangle$ vanishes in ground state, the Root Mean Square of the field $\langle E^2\rangle$ is not zero. Similarly the $\langle B^2\rangle$ is not zero. Therefore the electromagnetic field energy $\langle E^2\rangle+\langle B^2\rangle$ is not equal zero. A detailed theoretical calculation tells that EM energy in each mode of oscillation with frequency ω is 0.5 $\hbar\omega$, which equals one half of amount energy that would be present if a single “real” photon of that mode were present. Adding up 0.5 $\hbar\omega$ for each possible states of the electromagnetic field result in a very large number for the vacuum energy E_0 in a quantum vacuum

$$E_0 = \sum_i \frac{1}{2}\hbar\omega_i. \tag{1.153}$$

The resulting vacuum energy E_0 is *infinity* unless a high frequency cut off is applied.

Inserting surfaces into the vacuum causes the states of the EM field to change. This change in the states takes place because the EM field must meet the appropriate boundary conditions at each surface. The surfaces alter the modes of oscillation and therefore alter the energy density of the lowest state of the EM field. In actual practice the change in E_0 is

$$\Delta E_0 = E_0 - E_S, \tag{1.154}$$

where E_0 is the energy in empty space and E_S is the energy in space with surfaces, i.e.

$$\Delta E_0 = \frac{1}{2}\sum_n^{\text{empty space}} \hbar\omega_n - \frac{1}{2}\sum_i^{\text{surface present}} \hbar\omega_i. \tag{1.155}$$

As an example let us consider a hollow conducting rectangular cavity with sides a_1, a_2, a_3. In this case for uncharged parallel plates with an area A the attractive force between the plates is,

$$F_{att} = -\frac{\pi^2 \hbar c}{240 d^4} A, \tag{1.156}$$

where d is the distance between plates. The force F_{att} is called the parallel plate Casimir force, which was measured in three different experiments.

Recent calculations show that for conductive rectangular cavities the vacuum forces on a given face can be repulsive (positive), attractive (negative) or zero depending on the ratio of the sides.

The first measurement of repulsive Casimir force was performed by Maclay. For a distance of separation of $d \sim 0.1$ μm the repulsive force is of the order of 0.5 μN (micronewton) – for cavity geometry. In March 2001, scientist at Lucent Technology used the attractive parallel plate Casimir force to actuate a MEMS torsion device. Other MEMS (MicroElectroMechanical System) have been also proposed.

Standard Klein – Gordon equation is expressed as:

$$\frac{1}{c^2}\frac{\partial^2 \Psi}{\partial t^2} - \frac{\partial^2 \Psi}{\partial x^2} + \frac{m^2 c^2}{\hbar^2}\Psi = 0. \tag{1.157}$$

In equation (1.185) Ψ is the relativistic wave function for particle with mass m, c is the speed of light and $\hbar$ is Planck constant. In case of massless particles $m = 0$ and Eq. (1.185) is the Maxwell equation for photons. As was shown by Pauli and Weiskopf, the Klein – Gordon equation describes spin, – 0 bosons, because relativistic quantum mechanical equation had to allow for creation and annihilation of particles.

In the monograph by J. Marciak – Kozłowska and M. Kozłowski [1.8] the generalized Klein – Gordon thermal equation was developed

$$\frac{1}{\upsilon^2}\frac{\partial^2 T}{\partial t^2} - \nabla^2 T + \frac{m}{\hbar}\frac{\partial T}{\partial t} + \frac{2Vm}{\hbar^2} = 0. \tag{1.158}$$

In Eq. (1.158) T denotes temperature of the medium and υ is the velocity of the temperature signal in the medium. When we extract the highly oscillating part of the temperature field,

$$T = e^{-\frac{t\omega}{2}} u(x,t), \tag{1.159}$$

where $\omega = \tau^{-1}$, and τ is the relaxation time, we obtain from Eq. (1.155) (1D case)

$$\frac{1}{\upsilon^2}\frac{\partial^2 u}{\partial t^2} - \frac{\partial^2 u}{\partial x^2} + q^2 u(x,t) = 0, \tag{1.160}$$

where

$$q^2 = \frac{2Vm}{\hbar^2} - \left(\frac{m\upsilon}{2\hbar}\right)^2. \tag{1.161}$$

When $q^2 > 0$ equation (1.160) is of the form of the Klein – Gordon equation in the potential field $V(x, t)$. For $q^2 < 0$ Eq. (1.160) is the modified Klein – Gordon equation.

Considering the existence of the attosecond laser with $\Delta t = 1$ as $= 10^{-18}$s, Eq. (1.160) describes the heat signaling for thermal energy transport induced by ultra-short laser pulses. In the subsequent we will consider the heat transport when V is the Casimir potential. For attractive Casimir force, $V < 0$, $q^2 < 0$ (formula (1.161)) and equation (1.160) is the modified K-G equation. For repulsive Casimir force $V > 0$ and q^2 can be positive or negative.

References

[1.1] Corkum, P. B.; et al. *Phys. Rev. Lett.* 1998, *61*, 2886.

[1.2] Clemens, B. M.; et al. *Phys.* 1988, *B37*, 1085.

[1.3] Kozlowski, M.; Marciak – Kozlowska, J. *From Quarks to Bulk Matter*; Hadronic Press: Palm Harbor, FL, 2001.

[1.4] Dingle, R. H. *Proc. Phys. Soc.* 1952, *A65*, 374.

[1.5] Ward, J. C.; Wilks, J. *Philos. Mag.* 1951, *42*, 314.

[1.6] London, F. *Superfluids*; J. Wiley and Sons: New York, NY, 1954; Vol. 2.

[1.7] De Witt – Morette, C.; See Kit Fong. *Phys. Rev. Lett.* 1989, *19*, 2201.
[1.8] Kozlowski, M.; Marciak – Kozlowska, J. *Thermal Processes Using Attosecond Laser Pulses*; Optical Science 121; Springer: New York, NY, 2006.

Chapter 2

HEAVISIDE EQUATION FOR HYPERBOLIC TRANSPORT PROCESSES

2.1. BOLTZMANN EQUATION FOR HEAT TRANSPORT INDUCED BY ULTRA-SHORT LASER PULSES

Recently it has been shown that after optical excitation by femtosecond pulse establishment of an electron temperature by e-e scattering takes place on a few hundred femtosecond time scale in both bulk and nanostructured noble materials [2.1 – 2.4]. In noble metal clusters the electron thermalization time (relaxation time) is of the order of 200 fs [2.3, 2.4]. This relaxation time is much larger than the duration of the now available femtosecond optical pulses offering the unique possibility of analyzing the properties of a thermal quasi-free electron gas [2.5]. In paper [2.5] using a two color femtosecond pump-probe laser technique the ultrafast energy exchanges of a nonequilibrium electron gas was investigated. When the duration of the laser pulse, 25 fs, in paper [2.5], is shorter than the relaxation time the parabolic Fourier equation cannot be used [2.6, 2.7]. Instead, the new hyperbolic quantum heat transport equation is the valid equation [2.8]. The quantum heat transport equation is the wave damped equation for heat phenomena on the femtosecond scale.

Wave is an organized propagating imbalance [2.9]. Some phenomena seem to be clearly diffusive, with no wave-like implications, heat for instance. That was consistent with experiments at the late century, but not any longer. As far back as the 1960s ballistic (wave-like) heat pulses were observed at low temperatures [2.9]. The idea was that heat is just the manifestation of microscopic motion. Computing the classical resonant frequencies of atoms or molecules in a lattice gives numbers of the order of 10^{13} Hz, that is in the infrared, so when molecules jiggle they give off heat. These lattice vibrations are called phonons. Phonons have both wave-like and particle-like aspects. Lattice vibrations are responsible for the transport of heat, and we know that is a diffusive phenomenon, described by the Fourier equation. However, if the lattice is cooled to near absolute zero, the mean free scattering of the phonons becomes comparable to the macroscopic size of the sample. When this happens, lattice vibrations no longer behave diffusively but are actually wave-like or thermal wave. By controlling the temperature of a sample, one can control the extent to which heat is ballistic (thermal wave) or diffusive. In essence if a heat pulse is launched into sample (by the laser pulse interaction) and if the phonons can get across the sample without scattering, they will propagate as thermal waves.

The extent to which the motion of quasiparticles (phonons) or particles is ballistic, is described by the value of the relaxation time, τ. For ballistic (wave-like) motion, $\tau \to \infty$. The equation which is generalization of the Fourier equation (in which $\tau \to \infty$) is the Heaviside equation [2.6] for thermal processes:

$$\tau\frac{\partial^2 T}{\partial t^2}+\frac{\partial T}{\partial t}=D\nabla^2 T. \tag{2.1}$$

For very short relaxation time, $\tau \to 0$ we obtain from equation (2.1) the Fourier equation

$$\frac{\partial T}{\partial t}=D\nabla^2 T \tag{2.2}$$

and for $\tau \to \infty$ we obtain from formula (2.1), the ballistic ≡ thermal wave motion:

$$\frac{1}{\upsilon^2}\frac{\partial^2 T}{\partial t^2}=\nabla^2 T. \tag{2.3}$$

In the set of papers [2.6, 2.7, 2.8] the quantum generalization of the Heaviside equation was obtained and solved:

$$\frac{1}{\upsilon^2}\frac{\partial^2 T}{\partial t^2}+\frac{m}{\hbar}\frac{\partial T}{\partial t}=\frac{\partial^2 T}{\partial t^2}, \tag{2.4}$$

where $\upsilon = \alpha c$ and is the fine structure constant, c is the vacuum light velocity. In formula (2.4) m is the *heaton* mass [2.3]. *Heaton* energy is equal

$$E_h = m\alpha^2 c^2. \tag{2.5}$$

In paper [2.6] Heaviside equation was obtained for the fermionic gases (electrons, nucleons, quarks). In this chapter the Heaviside equation will be obtained for particles with mass m, where m is the mass of the fermion or boson. Moreover besides the elastic scattering of the particles, the creation and absorption of the heat carriers will be discussed. The new form of the discrete Heaviside equation will be obtained as the result of the discretization of the one-dimensional Boltzmann equation. The solution of the discrete Boltzmann equation will be obtained for Cauchy boundary conditions, initialed by ultra-short laser pulses, i.e. for $\Delta t \le \tau$, the relaxation time.

Let us consider the one-dimensional rod (strand) which can transport “particles” – heat carriers. These particles however may move only to the right or to the left on the rod. Moving particles may collide with the fixed scatter centra, barriers, dislocations) the probabilities of such collisions and their expected results being specified. All particles will be of the same kind, with the same energy and other physical specifications distinguishable only by their direction.

Let us define:

$u(z,t)$ = expected density of particles at z and at time t moving to the right,
$v(z,t)$ = expected density of particles at z and at time t moving to the left.
Furthermore, let

$\delta(z)$ = probability of collision occurring between a fixed scattering centrum and a particle moving between z and $z+\Delta$.

Suppose that a collision might result in the disappearance of the moving particle without new particle appearing. Such a phenomenon is called *absorption*. Or the moving particle may be reversed in direction or back-scattered. We shall agreeing that in each collision at z an expected total of $F(z)$ particles arises moving in the direction of the original particle, $B(z)$ arise going in the opposite direction.

The expected total number of right-moving particles in $z_1 \le z \le z_2$ at time t is

$$\int_{z_1}^{z_2} u(z,t)dz, \tag{2.6}$$

while the total number of particles passing z to the right in the time interval $t_1 \le t \le t_2$ is

$$w\int_{t_1}^{t_2} u(z,t)dt, \tag{2.7}$$

where w is the particles speed.

Consider the particle moving to the right and passing $z+\Delta$ in the time interval $t_1+\frac{\Delta}{w} \le t \le t_2+\frac{\Delta}{w}$:

$$w\int_{t_1+\Delta/w}^{t_2+\Delta/w} u(z+\Delta,t')dt' = w\int_{t_1}^{t_2} u\left(z+\Delta,t'+\frac{\Delta}{w}\right)dt'. \tag{2.8}$$

These can arise from particles which passed z in the time interval $t_1 \le t \le t_2$ and came through ($z, z+\Delta$) without collision

$$w\int_{t_1}^{t_2}(1-\Delta\delta(z,t'))u(z,t')dt' \tag{2.9}$$

plus contributions from collisions in the interval $(z, z+\Delta)$. The right-moving particles interacting in $(z, z+\Delta)$ produce in the time t_1 to t_2,

$$w\int_{t_1}^{t_2}\Delta\delta(z,t')F(z,t')u(z,t')dt' \tag{2.10}$$

particles to the right, while the left moving ones give:

$$w\int_{t_1}^{t_2}\Delta\delta(z,t')B(z,t')v(z,t')dt'. \tag{2.11}$$

Thus

$$w\int_{t_1}^{t_2}u\left(z+\Delta,t'+\frac{\Delta}{w}\right)dt' = w\int_{t_1}^{t_2}u(z,t')dt' + w\Delta\int_{t_1}^{t_2}\delta(z,t')(F(z,t')-1)u(z,t')dt' \\ +w\Delta\int_{t_1}^{t_2}\delta(z,t')B(z,t')v(z,t')dt'. \tag{2.12}$$

Now, we can write:

$$u\left(z+\Delta,t'+\frac{\Delta}{w}\right) = u(z,t') + \left(\frac{\partial u}{\partial z}(z,t') + \frac{1}{w}\frac{\partial u}{\partial t}(z,t')\right)\Delta \tag{2.13}$$

to get

$$\int_{t_1}^{t_2}\left(\frac{\partial u}{\partial z}(z,t')\right)+\frac{1}{w}\frac{\partial u}{\partial t}(z,t')dt' = \int_{t_1}^{t_2}\delta(z,t')((F(z,t')-1)u(z,t')+B(z,t')v(z,t'))dt'. \tag{2.14}$$

On letting $\Delta \to 0$ and differentiating with respect to t_2 we find

$$\frac{\partial u}{\partial z}+\frac{1}{w}\frac{\partial u}{\partial t} = \delta(z,t)(F(z,t)-1)u(z,t)+\delta(z,t)B(z,t)v(z,t). \tag{2.15}$$

In a like manner

$$-\frac{\partial v}{\partial z}+\frac{1}{w}\frac{\partial v}{\partial t}=\delta(z,t)B(z,t)u(z,t)+\delta(z,t)(F(z,t)-1)v(z,t). \tag{2.16}$$

The system of partial differential equations of hyperbolic type (2.15, 2.16) is the Boltzmann equation for one dimensional transport phenomena [2.10].

Let us define the total density for heat carriers, $\rho(z,t)$

$$\rho(z,t)=u(z,t)+v(z,t) \tag{2.17}$$

and density of heat current

$$j(z,t)=w(u(z,t)-v(z,t)). \tag{2.18}$$

Considering equations (2.15 – 2.18) one obtains

$$\frac{\partial \rho}{\partial z}+\frac{1}{w^2}\frac{\partial j}{\partial t}=\delta(z,t)u(z,t)(F(z,t)-B(z,t)-1)+\delta(z,t)v(z,t)(B(z,t)-F(z,t)+1). \tag{2.19}$$

Equation (2.19) can be written as

$$\frac{\partial \rho}{\partial z}+\frac{1}{w^2}\frac{\partial j}{\partial t}=\frac{\delta(z,t)(F(z,t)-B(z,t)-1)j}{w} \tag{2.20}$$

or

$$j=\frac{w}{\delta(z,t)(F(z,t)-B(z,t)-1)}\frac{\partial \rho}{\partial z}+\frac{1}{w\delta(z,t)(F(z,t)-B(z,t)-1)}\frac{\partial j}{\partial t}. \tag{2.21}$$

Denoting, D, diffusion coefficient

$$D=-\frac{w}{\delta(z,t)(F(z,t)-B(z,t)-1)}$$

and τ, relaxation time

$$\tau=\frac{1}{w\delta(z,t)(1-F(z,t)-B(z,t))} \tag{2.22}$$

equation (2.21) takes the form

$$j = -D\frac{\partial\rho}{\partial z} - \tau\frac{\partial j}{\partial t}. \tag{2.23}$$

Equation (2.23) is the Cattaneo's type equation and is the generalization of the Fourier equation. Now in a like manner we obtain from equation (2.15 – 2.18)

$$\begin{aligned}\frac{1}{w}\frac{\partial j}{\partial z} + \frac{1}{w}\frac{\partial\rho}{\partial t} &= \delta(z,t)u(z,t)(F(z,t) - 1 + B(z,t)) \\ &+\delta(z,t)v(z,t)(B(z,t) + F(z,t) - 1))\end{aligned} \tag{2.24}$$

Or

$$\frac{\partial j}{\partial z} + \frac{\partial\rho}{\partial t} = 0. \tag{2.25}$$

Equation (2.25) describes the conservation of energy in the transport processes.

Considering equations (2.23) and (2.25) for the constant D and τ the hyperbolic Heaviside equation is obtained:

$$\tau\frac{\partial^2\rho}{\partial t^2} + \frac{\partial\rho}{\partial t} = D\frac{\partial^2\rho}{\partial z^2}. \tag{2.26}$$

In the case of the *heaton* gas with temperature $T(z,t)$ equation (2.26) has the form [2.6 – 2.8]

$$\tau\frac{\partial^2 T}{\partial t^2} + \frac{\partial T}{\partial t} = D\frac{\partial^2 T}{\partial z^2},$$

where τ is the relaxation time for the thermal processes.

In the stationary state transport phenomena $dF(z,t)/dt = dB(z,t)dt = 0$ and $d\delta(z,t)/dt = 0$. In that case we denote $F(z,t) = F(z) = B(z,t) = B(z) = k(z)$ and equation (2.10) and (2.11) can be written as

$$\begin{aligned}\frac{du}{dz} &= \delta(z)(k-1)u(z) + \delta(z)kv(z), \\ -\frac{dv}{dz} &= \delta(z)k(z)u(z) + \delta(z)(k(z)-1)v(z)\end{aligned} \tag{2.27}$$

with diffusion coefficient

$$D = \frac{w}{\delta(z)} \tag{2.28}$$

and relaxation time

$$\tau(z) = \frac{1}{w\delta(z)(1-2k(z))}. \tag{2.29}$$

The system of equations (2.27) can be written as

$$\frac{d^2u}{dz^2} - \frac{\frac{d}{dz}(\delta k)}{\delta k}\frac{du}{dz} + u\left[\delta^2(2k-1) + \frac{d\delta}{dz}(1-k) + \frac{\delta(k-1)}{\delta k}\frac{d(\delta k)}{dz}\right] = 0, \tag{2.30}$$

$$\frac{du}{dz} = \delta(k-1)u + \delta k v(z). \tag{2.31}$$

Equation (2.30) after differentiation has the form

$$\frac{d^2u}{dz^2} + f(z)\frac{du}{dz} + g(z)u(z) = 0, \tag{2.32}$$

where

$$\begin{aligned} f(z) &= -\frac{1}{\delta}\left(\frac{\delta}{k}\frac{dk}{dz} + \frac{d\delta}{dz}\right), \\ g(z) &= \delta^2(z)(2k-1) - \frac{\delta}{k}\frac{dk}{dz}. \end{aligned} \tag{2.33}$$

For the constant absorption rate we put

$$k(z) = k = \text{constant} \neq \frac{1}{2}.$$

In that case

$$\begin{aligned} f(z) &= -\frac{1}{\delta}\frac{d\delta}{dz}, \\ g(z) &= \delta^2(z)(zk-1). \end{aligned} \tag{2.34}$$

With functions $f(z)$ and $g(z)$ the general solution of the equation (2.30) has the form

$$u(z) = C_1 e^{(1-2k)^{1/2}\int \delta dz} + C_2 e^{-(1-2k)^{1/2}\int \delta dz}. \tag{2.35}$$

In the subsequent we will consider the solution of the equation (2.32) with $f(z)$ and $g(z)$ described by (2.34) for Cauchy condition:

$$u(0) = q, \quad v(a) = 0. \tag{2.36}$$

Boundary condition (2.36) describes the generation of the heat carriers (by illuminating the left end of the strand with laser pulses) with velocity q heat carrier per second.

The solution has the form:

$$u(z) = \frac{2qe^{[f(0)-f(a)]}}{1+\beta e^{2[f(0)-f(a)]}}\left[\frac{(1-2k)^{\frac{1}{2}}}{(1-2k)^{\frac{1}{2}}-(k-1)}\cosh[f(x)-f(a)] + \frac{k-1}{(1-2k)^{\frac{1}{2}}-(k-1)}\sinh[f(x)-f(a)]\right] \tag{2.37}$$

$$u(z) = \frac{2qe^{(f(0)-f(a))}}{1+\beta e^{2[f(0)-f(a)]}}\left[\frac{(1-2k)^{\frac{1}{2}}+(k-1)}{k}\sinh[f(x)-f(a)]\right],$$

where

$$\begin{aligned}
f(z) &= (1-2k)^{\frac{1}{2}}\int \delta dz, \\
f(0) &= (1-2k)^{\frac{1}{2}}\left[\int \delta dz\right]_0, \\
f(a) &= (1-2k)^{\frac{1}{2}}\left[\int \delta dz\right]_a, \\
\beta &= \frac{(1-2k)^{\frac{1}{2}}+(k-1)}{(1-2k)^{\frac{1}{2}}-(k-1)}.
\end{aligned} \tag{2.38}$$

Considering formulae (2.17), (2.18) and (2.37) we obtain for the density, $\rho(z)$ and current density $j(z)$.

$$j(z)=\frac{2qwe^{[f(0)-f(a)]}}{1+\beta e^{2[f(0)-f(a)]}}\left[\begin{array}{l}\dfrac{(1-2k)^{\frac{1}{2}}}{(1-2k)^{\frac{1}{2}}-(k-1)}\cosh[f(z)-f(a)]\\-\dfrac{1-2k}{(1-2k)^{\frac{1}{2}}-(k-1)}\sinh[f(z)-f(a)]\end{array}\right] \tag{2.39}$$

and

$$q=\frac{2qe^{[f(0)-f(a)]}}{1+\beta e^{2[f(0)-f(a)]}}\left[\begin{array}{l}\dfrac{(1-2k)^{\frac{1}{2}}}{(1-2k)^{\frac{1}{2}}-(k-1)}\cosh[f(z)-f(a)]\\-\dfrac{1}{(1-2k)^{\frac{1}{2}}-(k-1)}\sinh[f(z)-f(a)]\end{array}\right]. \tag{2.40}$$

Equations (2.39) and (2.40) fulfill the generalized Fourier relation

$$j=-\frac{w}{\delta(z)}\frac{\partial\rho}{\partial z},\qquad D=\frac{W}{\delta(z)}, \tag{2.41}$$

where D denotes the diffusion coefficient.

Analogously we define the generalized diffusion velocity $\upsilon_D(z)$

$$\upsilon_D(z)=\frac{j(z)}{n(z)}=\frac{w(1-2k)^{\frac{1}{2}}\left[\cosh[f(z)-f(a)]-(1-2k)^{\frac{1}{2}}\sinh[f(x)-f(a)]\right]}{(1-2k)^{\frac{1}{2}}\cosh[f(x)-f(a)]-\sinh[f(x)-f(a)]}. \tag{2.42}$$

Assuming constant cross section for heat carriers scattering $\delta(z)=\delta_o$ we obtain from formula (2.38)

$$\begin{aligned}&f(z)=(1-2k)^{\frac{1}{2}}z,\\&f(0)=0,\\&f(a)=(1-2k)^{\frac{1}{2}}a\end{aligned} \tag{2.43}$$

and for density $\rho(z)$ and current density $j(z)$

$$j(z)=\frac{2qwe^{-(1-2k)^{\frac{1}{2}}a\delta}}{1+\beta e^{-(1-2k)^{\frac{1}{2}}a\delta}}\left[\frac{(1-2k)^{\frac{1}{2}}}{(1-2k)^{\frac{1}{2}}-(k-1)}\cosh\left[(2k-1)^{\frac{1}{2}}(x-a)\delta\right]\right.$$
$$\left.-\frac{(1-2k)}{(1-2k)^{\frac{1}{2}}-(k-1)}\sinh\left[(2k-1)^{\frac{1}{2}}(x-a)\delta\right]\right], \tag{2.44}$$

$$\rho(z)=\frac{2qe^{-(1-2k)^{\frac{1}{2}}a\delta}}{1+\beta e^{-(1-2k)^{\frac{1}{2}}a\delta}}\left[\frac{(1-2k)^{\frac{1}{2}}}{(1-2k)^{\frac{1}{2}}-(k-1)}\cosh\left[(2k-1)^{\frac{1}{2}\delta}(x-a)\right]\right.$$
$$\left.-\frac{1}{(1-2k)^{\frac{1}{2}}-(k-1)}\sinh\left[(2k-1)^{\frac{1}{2}}(x-a)\delta\right]\right]. \tag{2.45}$$

We define Fourier's diffusion velocity $\upsilon_F(z)$ and diffusion length, L

$$\upsilon_F=\left(\frac{D}{\tau}\right)^{\frac{1}{2}}, \qquad L=\upsilon_F\tau. \tag{2.46}$$

Considering formulae (2.28) and (2.29) one obtains

$$\upsilon_F(z)=w(1-2k)^{\frac{1}{2}},$$
$$L=\frac{1}{\delta(1-2k)^{\frac{1}{2}}}=\frac{\lambda_{\mathrm{mfp}}}{(1\text{-}2\mathrm{k})^{\frac{1}{2}}}, \tag{2.47}$$

where λ_{mfp} denotes the mean free path for heat carriers.

Considering formulae (2.44), (2.45), (2.46), (2.47) one obtains

$$j(z)=\frac{2qwe^{-\frac{a}{L}}}{1+\beta e^{-\frac{a}{L}}}\left[\frac{(1-2k)^{\frac{1}{2}}}{(1-2k)^{\frac{1}{2}}-(k-1)}\cosh\left[\frac{(x-a)}{L}\right]\right.$$
$$\left.-\frac{(1-2k)}{(1-2k)^{\frac{1}{2}}-(k-1)}\sinh\left[\frac{x-a}{L}\right]\right], \tag{2.48}$$

$$\rho(z)=\frac{2qe^{-\frac{a}{L}}}{1+\beta e^{-\frac{a}{L}}}\left[\frac{(1-2k)^{\frac{1}{2}}}{(1-2k)^{\frac{1}{2}}-1}\cosh\left[\frac{x-a}{L}\right]\right.$$
$$\left.-\frac{1}{(1-2k)^{\frac{1}{2}}-(k-1)}\sinh\left[\frac{x-a}{L}\right]\right]. \tag{2.49}$$

For the length of strand, $a >> L$ the both solutions Fourier and Boltzmann equations overlap. For $a \leq L$ the Boltzmann equation gives the different description of the transport processes. In that case the solution of the Boltzmann equation depends strongly on the scatterings (k coefficient) of the carriers. Recently [2.11], the heat conduction in one-dimensional system is actively investigated. As was discussed in paper [2.11] the dependence of density current on L can be described by the general formula

$$j \sim L^{\alpha},$$

where α can be positive or negative. The same conclusion can be drawn from the calculation presented in our paper. In this calculation coefficient α depends on the scattering cross section for the heat carriers.

2.2. Universal Relaxation Time for Electron and Nucleon Gases

The differential equations of thermal energy transfer should be hyperbolic so as to exclude action at distance; yet the equations of irreversible thermodynamics – those of Navier – Stokes and Fourier are parabolic.

In the book [2.12] the new hyperbolic non – Fourier equation for heat transport was formulated and solved.

The excitation of matter on the quark nuclear and atomic level leads to transfer of energy. The response of the chunk of matter (nucleus, atom) is governed by the relaxation time.

In this paragraph we develop the general, universal definition of the relaxation time, which depends on coupling constants for electromagnetic or strong interaction.

It occurs that the general formula for the relaxation time can be written as

$$\tau_i=\frac{\hbar}{m_i(\alpha_i c)^2} \tag{2.50}$$

where m_i is the heat carrier mass, $\alpha_i = \left(i = e, {}^1\!/_{137}, i = N, {}^{m_\pi}\!/_{m_n}\right)$ is coupling constant for electromagnetic and strong interaction, c is the vacuum light speed. As the c is the maximal velocity all relaxation time fulfils the inequality

$$\tau > \tau_i \tag{2.51}$$

Consequently τ_i is the minimal universal relaxation time.

Dynamical processes are commonly investigated using laser pump-probe experiments with a pump pulse exciting the system of interest and a second probe pulse tracking is temporal evolution. As the time resolution attainable in such experiments depends on the temporal definition of the laser pulse, pulse compression to the attosecond domain is a recent promising development.

After the standards of time and space were defined the laws of classical physics relating such parameters as distance, time, velocity, temperature are assumed to be independent of accuracy with which these parameters can be measured. It should be noted that this assumption does not enter explicitly into the formulation of classical physics. It implies that together with the assumption of existence of an object and really independently of any measurements (in classical physics) it was tacitly assumed that *there was a possibility of an unlimited increase in accuracy of measurements.* Bearing in mind the "atomicity" of time i.e. considering the smallest time period, the Planck time, the above statement is obviously not true. Attosecond laser pulses we are at the limit of laser time resolution.

With attosecond laser pulses belong to a new Nano – World where size becomes comparable to atomic dimensions, where transport phenomena follow different laws from that in the macro world. This first stage of miniaturization, from 10^{-3} m to 10^{-6} m is over and the new one, from 10^{-6} m to 10^{-9} m just beginning. The Nano – World is a quantum world with all the predicable and non-predicable (yet) features.

In paper [2.13] the relativistic hyperbolic transport equation was developed:

$$\frac{1}{\upsilon^2}\frac{\partial^2 T}{\partial t^2} + \frac{m_0\gamma}{\hbar}\frac{\partial T}{\partial t} = \nabla^2 T. \tag{2.52}$$

In equation (2.52) υ is the velocity of heat waves, m_0 is the mass of heat carrier and γ – the Lorentz factor, $\gamma = \left(1 - \frac{\upsilon^2}{c^2}\right)^{-1/2}$. As was shown in paper [2.13] the heat energy (*heaton temperature*) T_h can be defined as follows:

$$T_h = m_0\gamma\upsilon^2. \tag{2.53}$$

Considering that υ, the thermal wave velocity equals [2.13]

$$\upsilon = \alpha c, \tag{2.54}$$

where α is the coupling constant for the interactions which generate the *thermal wave* (α = 1/137 and α = 0.15 for electromagnetic and strong forces respectively), the *heaton temperature* is equal to

$$T_h = \frac{m_0 \alpha^2 c^2}{\sqrt{1-\alpha^2}}. \tag{2.55}$$

Based on equation (2.55) one concludes that the *heaton temperature* is a linear function of the mass m_0 of the heat carrier. It is interesting to observe that the proportionality of T_h and the heat carrier mass m_0 was observed for the first time in ultrahigh energy heavy ion reactions measured at CERN [2.14]. In paper [2.14] it was shown that the temperature of pions, kaons and protons produced in Pb+Pb, S+S reactions are proportional to the mass of particles. Recently, at Rutherford Appleton Laboratory (RAL), the VULCAN LASER was used to produce the elementary particles: electrons and pions [2.15].

When the external force is present $F(x,t)$ the forced damped heat transport is obtained [2.13] (in one dimensional case):

$$\frac{1}{\upsilon^2}\frac{\partial^2 T}{\partial t^2} + \frac{m_0\gamma}{\hbar}\frac{\partial T}{\partial t} + \frac{2Vm_0\gamma}{\hbar^2}T - \frac{\partial^2 T}{\partial x^2} = F(x,t). \tag{2.56}$$

The hyperbolic relativistic quantum heat transport equation, (2.56), describes the forced motion of heat carriers which undergo scattering ($\frac{m_0\gamma}{\hbar}\frac{\partial T}{\partial t}$ term) and are influenced by the potential term ($\frac{2Vm_o\gamma}{\hbar^2}T$).

Equation (2.56) is the Proca thermal equation and can be written as [2.13]:

$$\left(\bar{\Box}^2 + \frac{2Vm_0\gamma}{\hbar^2}\right)T + \frac{m_0\gamma}{\hbar}\frac{\partial T}{\partial t} = F(x,t),$$
$$\bar{\Box}^2 = \frac{1}{\upsilon^2}\frac{\partial^2}{\partial t^2} - \frac{\partial^2}{\partial x^2}. \tag{2.57}$$

We seek the solution of equation (2.57) in the form

$$T(x,t) = e^{-t/2\tau}u(x,t), \tag{2.58}$$

where $\tau_i = \frac{\hbar}{m\upsilon^2}$ is the relaxation time. After substituting equation (2.58) in equation (2.57) we obtain a new equation

$$\left(\bar{\Box}^2+q^2\right)u(x,t)=e^{t/2\tau}F(x,t) \tag{2.59}$$

and

$$q^2=\frac{2Vm}{\hbar^2}-\left(\frac{m\upsilon}{2\hbar}\right)^2, \tag{2.60}$$

$$m=m_0\gamma\,. \tag{2.61}$$

In free space i.e. when $F(x,t)\to 0$ equation (2.59) reduces to

$$\left(\bar{\Box}^2+q^2\right)u(x,t)=0, \tag{2.62}$$

which is essentially the free Proca type equation.

The Proca equation describes the interaction of the laser pulse with the matter. As was shown in book [2.12] the quantization of the temperature field leads to the *heatons* – quanta of thermal energy with a mass $m_h=\dfrac{\hbar}{\tau\upsilon_h^2}$ [2.12], where τ is the relaxation time and υ_h is the finite velocity for heat propagation. For $\upsilon_h\to\infty$, i.e. for $c\to\infty$, $m_0\to 0$, it can be concluded that in non-relativistic approximation (c = infinite) the Proca equation is the diffusion equation for massless photons and heatons.

For the initial *Cauchy* condition:

$$u(x,0)=f(x),\qquad\qquad u_t(x,0)=g(x) \tag{2.63}$$

the solution of the Proca equation has the form (for $q^2>0$) [2.13]

$$\begin{aligned}
u(x,t)=&\frac{f(x-\upsilon t)+f(x+\upsilon t)}{2}\\
&+\frac{1}{2\upsilon}\int_{x-\upsilon t}^{x+\upsilon t}g(\varsigma)J_0\left[\sqrt{q^2\left(\upsilon^2t^2-(x-\varsigma)^2\right)}\right]d\varsigma\\
&-\frac{\sqrt{q^2}\,\upsilon t}{2}\int_{x-\upsilon t}^{x+\upsilon t}f(\varsigma)\frac{J_1\left[\sqrt{q^2\left(\upsilon^2t^2-(x-\varsigma)^2\right)}\right]}{\sqrt{\upsilon^2t^2-(x-\varsigma)^2}}d\varsigma\\
&+\frac{1}{2\upsilon}\int_0^t\int_{x-\upsilon(t-t')}^{x+\upsilon(t-t')}G(\varsigma,t')J_0\left[\sqrt{q^2\left(\upsilon^2(t-t')^2-(x-\varsigma)^2\right)}\right]dt'd\varsigma,
\end{aligned} \tag{2.64}$$

where $G = e^{t/2\tau} F(x,t)$.

When $q^2 < 0$ solution of Proca equation has the form:

$$
\begin{aligned}
u(x,t) &= \frac{f(x-\upsilon t) + f(x+\upsilon t)}{2} \\
&+ \frac{1}{2\upsilon} \int_{x-\upsilon t}^{x+\upsilon t} g(\varsigma) I_0 \left[\sqrt{-q^2 \left(\upsilon^2 t^2 - (x-\varsigma)^2 \right)} \right] d\varsigma \\
&- \frac{\sqrt{-q^2}\, \upsilon t}{2} \int_{x-\upsilon t}^{x+\upsilon t} f(\varsigma) \frac{I_1 \left[\sqrt{-q^2 \left(\upsilon^2 t^2 - (x-\varsigma)^2 \right)} \right]}{\sqrt{\upsilon^2 t^2 - (x-\varsigma)^2}} d\varsigma \\
&+ \frac{1}{2\upsilon} \int_0^t \int_{x-\upsilon(t-t')}^{x+\upsilon(t-t')} G(\varsigma,t') I_0 \left[\sqrt{-q^2 \left(\upsilon^2 (t-t')^2 - (x-\varsigma)^2 \right)} \right] dt' d\varsigma.
\end{aligned}
\tag{2.65}
$$

When $q^2 = 0$ equation (2.59) is the forced thermal equation

$$
\frac{1}{\upsilon^2} \frac{\partial^2 u}{\partial t^2} - \frac{\partial^2 u}{\partial x^2} = G(x,t). \tag{2.66}
$$

In this paragraph we developed the relativistic thermal transport equation for an attosecond laser pulse interaction with matter. It is shown that the equation obtained is the Proca equation, well known in relativistic electrodynamics for massive photons. As the *heatons* are massive particles the analogy is well founded.

2.3. Nonlinear Klein – Gordon Equation for Nanoscale Transport

The study of transport mechanisms at the nanoscale level is of great importance nowadays. Specifically, the nanoparticles and nanotubules have important physical applications for nano- and micro-scale technologies [2.12]. Many models have been developed in the simple picture of point-like particles. One possibility that has been considered in the literature is that of nonlinear Klein – Gordon system where the on-site potential is ratchet-like.

The development of the ultra-short laser pulses opens new possibilities in the study of the dynamics of the electrons in nanoscale systems: carbon nanotubes, nanoparticles. For attosecond laser pulses the duration of the pulse is shorter than the relaxation time. In that case the transport equations contain the second order partial derivative in time. The master equation is the Klein – Gordon equation.

In this paragraph we consider the non - linear Klein – Gordon equation for mass and thermal energy transport in nanoscale. Considering the results of the monograph [2.12] we develop the nonlinear Klein – Gordon equation for heat and mass transport in nanoscale. For ultrashort laser pulse the nonlinear Klein – Gordon equation is reduced to the nonlinear d'Alembert equation. In this chapter we find out the implicit solution of the nonlinear d'Alembert equation for heat transport on nanoscale. It will be shown that for ultra-short laser pulses the non-linear Klein – Gordon equation has the nonlinear traveling wave solution.

In monograph [2.12] it was shown that in the case of the ultra short laser pulses the heat transport is described by the Heaviside hyperbolic heat transport equation:

$$\tau\frac{\partial^2 T}{\partial t^2}+\frac{\partial T}{\partial t}=D\frac{\partial^2 T}{\partial x^2}, \tag{2.67}$$

where T denotes the temperature of the electron gas in nanoparticle, τ is the relaxation time, m is the electron mass and D is the thermal diffusion coefficient. The relaxation time τ is defined as:

$$\tau=\frac{\hbar}{m\upsilon^2},\quad \upsilon=\alpha c, \tag{2.68}$$

where υ is the thermal pulse propagation speed. For electromagnetic interaction when scatters are the relativistic electrons, τ = Thomson relaxation time

$$\tau=\frac{\hbar}{mc^2}. \tag{2.69}$$

Both parameters τ and υ completely characterize the thermal energy transport on the atomic scale and can be named as "atomic" relaxation time and "atomic" heat velocity.

In the following, starting with the atomic τ and υ we describe thermal relaxation processes in nanoparticles which consist of N light scatters. To that aim we use the Pauli-Heisenberg inequality [2.16]:

$$\Delta r\Delta p\geq N^{\frac{1}{3}}\hbar, \tag{2.70}$$

where r denotes the radius of the nanoparticle and p is the momentum of energy carriers.

According to formula (2.70) we recalculate the relaxation time τ for nanoparticle consisting N electrons:

$$\hbar^N(N)=N^{\frac{1}{3}}\hbar, \tag{2.71}$$

$$\tau^N=N\tau. \tag{2.72}$$

Formula (2.72) describes the scaling of the relaxation time for *N fermion* systems. With formulae (2.70) and (2.71) the heat transport equation takes the form:

$$\tau^N \frac{\partial^2 T}{\partial t^2} + \frac{\partial T}{\partial t} = \frac{\hbar^{\frac{1}{3}}}{m} \frac{\partial^2 T}{\partial x^2} \tag{2.73}$$

and for mass transport:

$$\tau^N \frac{\partial^2 N}{\partial t^2} + \frac{\partial N}{\partial t} = \frac{N^{\frac{1}{3}}\hbar}{m} \frac{\partial^2 N}{\partial x^2}. \tag{2.74}$$

Equation (2.73) is linear damped Klein-Gordon equation, and was solved for nanotechnology systems in [2.12].

The nonlinearity of Eq. (2.74) opens new possibilities for the study of non-stationary stable processes in molecular nanostructures. Let us consider equation (2.74) in more details:

$$(N\tau) \frac{\partial^2 N}{\partial t^2} + \frac{\partial N}{\partial t} = \frac{N^{\frac{1}{3}}\hbar}{m} \frac{\partial^2 N}{\partial x^2}. \tag{2.75}$$

For $\Delta t < N\tau$ Eq (2.75) is the nonlinear d'Alembert equation

$$\frac{1}{\upsilon^2} \frac{\partial^2 N}{\partial t^2} = \frac{\partial^2 N}{\partial x^2} \tag{2.76}$$

with N dependent velocity:

$$\upsilon = \frac{1}{N^{\frac{1}{3}}} \alpha c. \tag{2.77}$$

Equation (2.76) can be written in more general form:

$$\frac{\partial^2 N}{\partial t^2} = \frac{\partial G(N)}{\partial x}, \tag{2.78}$$

where

$$G(N) = f(N) \frac{\partial N}{\partial x}. \tag{2.79}$$

The traveling wave solution of equation (2.78) has implicit form [2.17]:

$$\lambda^2 N(x,t) - \int G(N)dN = A(x+\lambda t) + B, \tag{2.80}$$

where A, B and λ are arbitrary constants.

2.4. Sub- and Supersonic Transport Processes

Recently it has become possible to produce MeV electrons with short-pulse multiteravat laser system [2.18]. The fast ignitor concept [2.19, 2.20] relevant to the inertial confinement fusion enhances the interest in this process. In an under-dense plasma, electrons and ions tend to be expelled from the focal spot by the ponderomotive pressure of an intense laser pulse, and the formed channel [2.21, 2.22] can act as a propagation guide for the laser beam. Depending on the quality of the laser beam, the cumulative effects of ponderomotive and relativistic self focusing [2.22] can significantly increase the laser intensity. For these laser pulses, the laser electric and magnetic fields reach few hundreds of GV/m and megagauss, respectively, and quiver velocity in the laser field is closed to the light speed. The component of the resulting Lorentz force $\left(e\vec{v}\times\vec{B}\right)$ accelerates electrons in the longitudinal direction, and energies of several tens of MeV can be achieved [2.23]. Recently the spectra of hot electrons (i.e. with energy in MeV region) were investigated. In paper [2.24] the interaction of 500 fs FWHM pulses with CH target was measured. The electrons with energy up to 20 MeV were observed. Moreover for electrons with energies higher than 5 MeV the change of electron temperature was observed: from 1 MeV (for energy of electrons $<$ 5 MeV) to 3 MeV (for energy of electrons $>$ 5 MeV). In this chapter the interaction of femtosecond laser pulse with electron plasma will be investigated. Within the theoretical framework of Heaviside temperature wave equation, the heating process of the plasma will be described. It will be shown that in vicinity of energy of 5 MeV the sound velocity in plasma reaches the value $\frac{1}{\sqrt{3}}c$ and is independent of the electron energy.

The mathematical form of the hyperbolic quantum heat transport was proposed in [2.25] and [2.26]. Under the absence of heat or mass sources the equations can be written as the Heaviside equations:

$$\frac{1}{\upsilon_\rho^2}\frac{\partial^2\rho}{\partial t^2} + \frac{1}{D_\rho}\frac{\partial\rho}{\partial t} = \frac{\partial^2\rho}{\partial x^2} \tag{2.81}$$

and

$$\frac{1}{\upsilon_T^2}\frac{\partial^2 T}{\partial t^2} + \frac{1}{D_T}\frac{\partial T}{\partial t} = \frac{\partial^2 T}{\partial x^2} \tag{2.82}$$

for mass and thermal energy transport respectively. The discussion of the properties of Eq. (2.81) was performed in [2.25] and Eq. (2.82) in [2.26]. In Eq. (2.81) υ_ρ is the velocity of density wave, D_ρ is the diffusion coefficient for mass transfer. In Eq. (2.82) υ_T is the velocity for thermal energy propagation and D_T is the thermal diffusion coefficient.

In the subsequent we will discuss the complex transport phenomena, i.e. diffusion and convection in the external field. The current density in the case when the diffusion and convection are taken into account can be written as:

$$j = -D_\rho \frac{\partial \rho}{\partial t} - \tau_\rho \frac{\partial j}{\partial t} + \rho V. \tag{2.83}$$

In equation (2.83) the first term describes the Fourier diffusion, the second term is the Maxwell – Cattaneo term and the third term describes the convection with velocity V. The continuity equation for the transport phenomena has the form:

$$\frac{\partial j}{\partial x} + \frac{\partial \rho}{\partial t} = 0. \tag{2.84}$$

Considering both equations (2.83) and (2.84) one obtains the transport equation:

$$\frac{\partial \rho}{\partial t} = -\tau_\rho \frac{\partial^2 \rho}{\partial t^2} + D_\rho \frac{\partial^2 \rho}{\partial x^2} - V\frac{\partial \rho}{\partial x}. \tag{2.85}$$

In equation (2.85) τ_ρ denotes the relaxation time for transport phenomena. Let us perform the Smoluchowski transformation for $\rho(x,t)$

$$\rho = \exp\left[\frac{Vx}{2D} - \frac{V^2 t}{4D}\right]\rho_1(x,t). \tag{2.86}$$

After substituting $\rho(x,t)$ formula (2.86) to equation (2.85) one obtains for $\rho_1(x,t)$:

$$\tau_\rho \frac{\partial^2 \rho_1}{\partial t^2} + \left(1 - \tau_\rho \frac{V_\rho^2}{2D_\rho}\right)\frac{\partial \rho_1}{\partial t} + \tau_\rho \frac{V_\rho^4}{16D_\rho^2}\rho_1 = D_\rho \frac{\partial^2 \rho_1}{\partial x^2}. \tag{2.87}$$

Considering that $D_\rho = \tau_\rho \upsilon_\rho^2$ equation (2.87) can be written as

$$\tau_\rho \frac{\partial^2 \rho_1}{\partial t^2} + \left(1 - \frac{V_\rho^2}{2\upsilon_\rho^2}\right)\frac{\partial \rho_1}{\partial t} + \frac{1}{16\tau_\rho}\frac{V_\rho^4}{\upsilon_\rho^4}\rho_1 = D_\rho \frac{\partial^2 \rho_1}{\partial x^2}. \tag{2.88}$$

In the same manner equation for the temperature field can be obtained:

$$\tau_T \frac{\partial^2 T_1}{\partial t^2} + \left(1 - \frac{V_T^2}{2\upsilon_T^2}\right)\frac{\partial T_1}{\partial t} + \frac{1}{16\tau_T}\frac{V_T^4}{\upsilon_T^4} T_1 = D_T \frac{\partial^2 T_1}{\partial x^2}. \qquad (2.89)$$

In equation (2.89) τ_T , D_T, V_T and υ_T are: relaxation time for heat transfer, diffusion coefficient, heat convection velocity and thermal wave velocity.

In this paragraph we will investigate the structure and solution of the equation (2.89). For the hyperbolic heat transport Eq. (2.89) we seek a solution of the form:

$$T_1(x,t) = e^{-\frac{t}{2\tau_T}} u(x,t). \qquad (2.90)$$

After substitution of Eq. (2.90) into Eq. (2.89) one obtains:

$$\tau_T \frac{\partial^2 u(x,t)}{\partial t^2} - D_T \frac{\partial^2 u(x,t)}{\partial x^2} + \left(-\frac{1}{4\tau_T} + \frac{V_T^2}{4D_T} + \tau_T \frac{V_T^4}{16D_T^2}\right) u(x,t) - \tau_T \frac{V_T^2}{2D_T}\frac{\partial u(x,t)}{\partial t} = 0. \qquad (2.91)$$

Considering that $D_T = \tau_T \upsilon_T^2$ Eq. (2.91) can be written as

$$\tau_T \frac{\partial^2 u}{\partial t^2} - \tau_T \upsilon_T^2 \frac{\partial^2 u}{\partial x^2} + \left(-\frac{1}{4\tau_T} + \frac{V_T^2}{4\tau_T \upsilon_T^2} + \tau_T \frac{V_T^4}{16\tau_T^2 \upsilon_T^4}\right) u(x,t) - \tau_T \frac{V_T^2}{2\upsilon_T^2}\frac{\partial u}{\partial t} = 0. \qquad (2.92)$$

After omitting the term $\frac{V_T^4}{\upsilon_T^4}$ in comparison to the term $\frac{V_T^2}{\upsilon_T^2}$ Eq. (2.92) takes the form:

$$\frac{\partial^2 u}{\partial t^2} - \upsilon_T^2 \frac{\partial^2 u}{\partial x^2} + \frac{1}{4\tau_T^2}\left(-1 + \frac{V_T^2}{\upsilon_T^2}\right) u(x,t) - \frac{V_T^2}{2\upsilon_T^2 \tau_T}\frac{\partial u}{\partial t} = 0. \qquad (2.93)$$

Considering that $\tau_T^{-2} >> \tau_T^{-1}$ one obtains from Eq. (2.39)

$$\frac{\partial^2 u}{\partial t^2} - \upsilon_T^2 \frac{\partial^2 u}{\partial x^2} + \frac{1}{4\tau_T^2}\left(-1 + \frac{V_T^2}{\upsilon_T^2}\right) u(x,t) = 0. \qquad (2.94)$$

Equation (2.94) is the master equation for heat transfer induced by ultra-short laser pulses, i.e. when $\Delta t \approx \tau_T$. In the following we will consider the Eq. (2.91) in the form:

$$\frac{\partial^2 u}{\partial t^2} - \upsilon_T^2 \frac{\partial^2 u}{\partial x^2} - q^2 u(x,t) = 0, \qquad (2.95)$$

where

$$q^2 = \frac{1}{4\tau_T^2}\left(\frac{V_T^2}{v_T^2} - 1\right). \tag{2.96}$$

In equation (2.96) the ratio

$$M_T = \frac{V_T}{v_T} = \frac{V_T}{v_S} \tag{2.97}$$

is the Mach number for thermal processes, for $v_T = v_S$ is the sound velocity in the gas of heat carriers [2.26].

In monograph [2.26] the structure of equation (2.95) was investigated. It was shown that for $q^2 < 0$, i.e. $V_T < v_S$, subsonic heat transfer is described by the modified telegrapher equation

$$\frac{1}{v_T^2}\frac{\partial^2 u}{\partial t^2} - \frac{\partial^2 u}{\partial x^2} + \frac{1}{4\tau_T^2 v_T^2}\left(\frac{V_T^2}{v_S^2} - 1\right)u(x,t) = 0. \tag{2.98}$$

For $q^2 > 0$, $V_T > v_S$, i.e. for supersonic case heat transport is described by Klein – Gordon equation:

$$\frac{1}{v_T^2}\frac{\partial^2 u}{\partial t^2} - \frac{\partial^2 u}{\partial x^2} + \frac{1}{4\tau_T^2 v_T^2}\left(\frac{V_T^2}{v_S^2} - 1\right)u(x,t) = 0. \tag{2.99}$$

The velocity of sound v_S depends on the temperature of the heat carriers. The general formula for sound velocity reads [2.27]:

$$v_S^2 = \left(zG - \frac{G}{z}\left(1 + \frac{5G}{z} - G^5\right)^{-1}\right)^{-1}. \tag{2.100}$$

In formula (2.100) $z = \dfrac{mc^2}{T}$ and G is of the form [2.27]:

$$G = \frac{K_3(z)}{K_2(z)}, \tag{2.101}$$

where c is the light velocity, m is the mass of heat carrier, T is the temperature of the gas and $K_3(z)$, $K_2(z)$ are modified Bessel functions of the second kind. For the initial conditions

$$u(x,0) = f(x), \qquad \left[\frac{\partial u(x,t)}{\partial t}\right]_{t=0} = F(x) \tag{2.102}$$

solution of the equation can be find in [2.26]:

$$u(x,t) = \frac{f(x-\upsilon_T t) + f(x+\upsilon_T t)}{2} + \frac{1}{2}\int_{x-\upsilon_T t}^{x+\upsilon_T t} \Phi(x,t,z)dz, \tag{2.103}$$

where

$$\Phi(x,t,z) = \frac{1}{\upsilon_T} F(z) J_0\left(\frac{\sqrt{q^2}}{\upsilon_T}\sqrt{(z-x)^2 - \upsilon_T^2 t^2}\right) + \sqrt{q^2}\, t f(z) \frac{J_0'\left(\frac{\sqrt{q^2}}{\upsilon_T}\sqrt{(z-x)^2 - \upsilon_T^2 t^2}\right)}{\sqrt{(z-x)^2 - \upsilon_T^2 t^2}}$$

and

$$q^2 = \frac{1}{4\tau_T^2}\left(\frac{V_T^2}{\upsilon_T^2} - 1\right).$$

The general equation for complex heat transfer: diffusion plus convection can be written as:

$$\frac{\partial T}{\partial t} = -\tau_T \frac{\partial^2 T}{\partial t^2} + D_T \frac{\partial^2 T}{\partial x^2} - V_T \frac{\partial T}{\partial x}. \tag{2.104}$$

Considering Eqs. (2.86), (2.90) and (2.103) the solution of equation (2.104) is

$$T(x,t) = \exp\left[\frac{V_T x}{2D} - \frac{V_T^2 t}{4D}\right] \exp\left[-\frac{t}{2\tau_T}\right] \cdot \left(\frac{f(x-\upsilon_T t) + f(x+\upsilon_T t)}{2} + \frac{1}{2}\int_{x-\upsilon_T t}^{x+\upsilon_T t} \Phi(x,t,z)dz\right). \tag{2.105}$$

2.5. Transport Processes in N – Dimensional Space – Time

The fact that we perceive the world to have three spatial dimensions is some-thing so familiar to our experience of its structure that we seldom pause to consider the direct influence this special property has upon the laws of physics. Yet some have done so and there have been many intriguing at-tempts to deduce the expediency or inevitability of a three-dimensional world from the general structure of the physical law themselves.

As earlier as 1917 P. Ehrenfest [2.28] pointed out that neither classical atoms nor planetary orbits can be stable in a space with $n > 3$ and traditional quantum atoms cannot be stable either [2.29]. As far as $n < 3$ is concerned, it has been argued [2.30] that organism would face insurmountable topological problem if $n = 2$: for instance two nerves cannot across. In the following we will conjecture that since $n = 2$ offers vastly less complexity that $n = 3$, worlds with $n < 3$ are just too simple and barren to contain observers. Since our Universe appears governed by the propagation of classical and quantum waves it is interesting elucidate the nature of the connection the properties of the wave equation and the spatial dimensions.

In this chapter we describe the partial differential equation (PDE) for the propagation of the thermal waves in n-dimensional space time. It is well known that for heat transport induced by ultra-short laser pulses (shorter than the relaxation time) the governing equation can be written as [2.26]

$$\frac{1}{\upsilon^2}\frac{\partial^2 T}{\partial t^2}+\frac{1}{D}\frac{\partial T}{\partial t}+\frac{2Vm}{\hbar^2}T=\nabla^2 T, \tag{2.106}$$

where T is the temperature, υ denotes the thermal wave propagation, m is the mass of heat carriers and V is the potential.

In monograph [2.26] the solution of the equation for one-dimensional case, $n = 1$ was obtained. In this paper we develop and solve the analog of the equation for n = natural numbers $n = 1, 2, \ldots$, separately for n = odd and n = even. The Huygens' principle for thermal wave will be discussed. It will be shown that for thermal waves only in odd dimensional space the waves propagate at exactly a fixed space velocity υ without "echoes" assuming the absence of walls (potentials) or inhomogeneities.

The three-dimensional heat transfer induced by ultra-short laser pulses in Cu_3Au alloy will be suggested.

In the following we consider the n-dimensional heat transfer phenomena described by the equation [2.26]

$$\frac{1}{\upsilon^2}\frac{\partial^2 T}{\partial t^2}+\frac{1}{D}\frac{\partial T}{\partial t}+\frac{2Vm}{\hbar^2}T=\nabla^2 T, \tag{2.107}$$

where temperature T is the function in the n-dimensional space

$$T = T(x_1, x_2, \ldots, x_n, t). \tag{2.108}$$

We seek solution of equation (2.107) in the form:

$$T(x_1, x_2, \ldots, x_n, t) = e^{-\frac{t}{2\tau}}u(x_1, x_2, \ldots, x_n, t). \tag{2.109}$$

After substitution of Eq. (2.109) to Eq. (2.107) one obtains

$$\frac{1}{\upsilon^2}\frac{\partial^2 u}{\partial t^2} - \nabla^2 u + q^2 u = 0, \tag{2.110}$$

where

$$q^2 = \frac{2Vm}{\hbar^2} - \left(\frac{m\upsilon}{2\hbar}\right)^2$$

for $D = \frac{\hbar}{m}$ [2.26].

We can define the distortionless thermal wave as the wave which preserves the shape in the field of the potential V. The condition for conserving the shape can be formulated as

$$q^2 = \frac{2Vm}{\hbar^2} - \left(\frac{m\upsilon}{2\hbar}\right)^2 = 0. \tag{2.111}$$

When Eq. (2.111) holds Eq. (2.110) has the form

$$\frac{1}{\upsilon^2}\frac{\partial^2 u}{\partial t^2} - \nabla^2 u = 0 \tag{2.112}$$

and condition (2.111) can be written as

$$V\tau \sim \hbar. \tag{2.113}$$

We conclude that in the presence of the potential energy V one can observe the undisturbed thermal wave only when the Heisenberg uncertainty relation (2.113) is fulfilled. The solution of the Eq. (2.112) for the n-odd can be found in [2.31]. First of all let us change the variables in Eq. (2.112)

$$\upsilon : t \to t', \quad x \to x', \quad u \to u'$$

and obtain

$$\frac{\partial^2 u'}{\partial t'^2} - \nabla^2 u' = 0. \tag{2.114}$$

for

$$\lim_{(x',t')\to(x^0,0)} u'(x',t') = g(x_0),$$

$$\lim_{(x',t')\to(x^0,0)} \frac{\partial u'(x',t')}{\partial t} = h(x_0) \tag{2.115}$$

the solution has the form [2.31]

$$u'(x',t') = \frac{1}{\gamma_n}\left[\left(\frac{\partial}{\partial t'}\right)\left(\frac{1}{t'}\frac{\partial}{\partial t'}\right)^{\frac{n-3}{2}}\left(t'^{n-2}\oint_{\partial B(x',t')} gdS\right) + \left(\frac{1}{t'}\frac{\partial}{\partial t'}\right)^{\frac{n-3}{2}}\left(t'^{n-2}\oint_{\partial B(x',t')} hdS\right)\right] \tag{2.116}$$

and $\gamma_n = 1\cdot 3\cdot 5\cdots(n-2)$.

For n – even the solution of equation (2.114) has the form [2.31].

$$u'(x',t') = \frac{1}{\gamma_n}\left[\begin{array}{l}\left(\frac{\partial}{\partial t'}\right)\left(\frac{1}{t'}\frac{\partial}{\partial t'}\right)^{\frac{n-2}{2}}\left(t'^{n}\oint_{\partial B(x',t')} \frac{g(y')dy'}{\left(t'^2-|y'-x'|^2\right)^{\frac{1}{2}}}\right)\\ +\left(\frac{1}{t'}\frac{\partial}{\partial t'}\right)^{\frac{n-2}{2}}\left(t'^{n}\oint_{\partial B(x',t')} \frac{h(y')dy'}{\left(t'^2-|y'-x'|^2\right)^{\frac{1}{2}}}\right)\end{array}\right] \tag{2.117}$$

and $\gamma_n = 2\cdot 4\cdots(n-2)\cdot n$. In formulae (2.116) and (2.117) $\oint$ denotes integral over n – space.

Considering formulae (2.119) and (2.117) we conclude that for n-odd the solution (2.116) is dependent on the value of functions h and g only on the hypersphere $\partial B(x',t')$. On the other hand for n-even the solution (2.117) is dependent on the values of the functions h and g on the full hyperball $B(x',t')$. In the other words for n-odd $n \geq 3$ the value of the initial functions h and g influences the solution (2.117) only on the surface of the cone $\{(y',t'), t'>0, |x'-y'| = t'\}$. For n = even the value of the functions g and h influences the solution on the full cone. It means that the thermal wave induced by the disturbance for n = odd have the well defined front. For n-even the wave influences space after the transmission of the front. This means that Huygens' principle is false for n-even. In conclusion: if we solve the wave equation in n-dimensions the signals propagate sharply (i.e. Huygens' principle is valid) only for dimensions n = 3, 5, 7, ... Thus three is the "best of all possible" dimensions, the smallest dimension in which signals propagate sharply.

The hyperbolic transport equation for heat transport (2.106)

$$\frac{1}{\upsilon_T^2}\frac{\partial^2 T}{\partial t^2} + \frac{1}{D_T}\frac{\partial T}{\partial t} + \frac{2Vm}{\hbar^2}T = \nabla^2 T \tag{2.118}$$

or mass transport

$$\frac{1}{\upsilon_\rho^2}\frac{\partial^2 \rho}{\partial t^2}+\frac{1}{D_\rho}\frac{\partial \rho}{\partial t}+\frac{2Vm}{\hbar^2}\rho=\nabla^2\rho \tag{2.119}$$

are the damped wave equations. For very short time period $\Delta t \sim \tau$ equations (2.118) and (2.119) can be written as

$$\frac{1}{\upsilon_{\rho,T}^2}\frac{\partial^2 u_{\rho,T}}{\partial t^2}-\nabla^2 u_{\rho,T}=0. \tag{2.120}$$

Eq. (2.120) is the generalization of equation (2.112). The solution of equation (2.120) in n- dimensional cases is described by formulae (2.116) and (2.117).

As was discussed in this paragraph only in 3-dimensional case the Huygens principle is fulfilled. It seems that in order to observe the thermal wave not disturbed by the "echoes" and with sharp front *the true three-dimensional experiment* must be performed. Moreover the experiment must be performed in the relaxation regime, i.e. for materials with relatively long relaxation time. The best candidate for "relaxation materials" is the Cu_3Au alloy [2.32]. As was shown in paper [2.32] the relaxation time is of the order of 10^4 s in the temperature range 650 – 660 K. For $t > 660$ K the abrupt increasing, up to $1.5 \cdot 10^5$ s (due to order → disorder transition) was observed.

References

[2.1] Sun, C. K.; et al. *Phys. Rev.* 1994, *B50*, 15337.
[2.2] Del Fatti, N.; et al. *Phys. Rev.* 2000, *B61*, 16956.
[2.3] Voisin, C.; et al. *Phys. Rev. Lett.* 2000, *85*, 2200.
[2.4] Del Fatti, N.; Vallée, F. *C. R. Acad. Sci.* 2002, *3*, 365.
[2.5] Guillon, C.; et al. *New Journal of Physics* 2003, *5*, 131.
[2.6] Kozlowski, M.; Marciak – Kozlowska, J. *Lasers in Engineering* 1998, *7*, 81.
[2.7] Kozlowski, M.; Marciak – Kozlowska, J. *Lasers in Engineering* 1999, *9*, 39.
[2.8] Kozlowski, M.; Marciak – Kozlowska, J. *Lasers in Engineering* 2000, *10*, 293.
[2.9] Wolfe, J. P. *Imaging Phonons Acoustic Wave Propagation in Solids;* Cambridge Univ. Press, 1998.
[2.10] Bellman, R.; Kalaba, R.; Wing, G. *J. Math. and Mech.* 1960, *9*, 933.
[2.11] Narayan, O.; Ramaswamy, S. *Phys. Rev. Lett.* 2002, *89*, 200601-1.
[2.12] Kozlowski, M.; Marciak – Kozlowska, J. *Thermal Processes Using Attosecond Laser Pulses*; Optical Science 121; Springer: New York, NY, 2006.
[2.13] Pelc, M.; Marciak – Kozlowska, J.; Kozlowski M. *Lasers in Engineering* 2005, *15*, 347.
[2.14] Bearden, I. G.; et al. *Phys. Rev. Lett.* 1997, *78*, 2080.
[2.15] Ledingham, K. W. D.; Norreys, P. A. *Contemporary Physics* 1999, *40*, 367.
[2.16] Ames, W. F.; et al. *Int. J. Nonlinear Mech.* 1981, *16*, 439.

[2.17] Polyanin, A. D.; Zaitsev, V. F. *Handbook of Nonlinear Partial Differential Equation*; Chapman and Hall/CRC, 2004.

[2.18] Blanchot, N.; et al. *Opt. Lett.* 1995, *20*, 395.

[2.19] Tabak, M.; et al. *Phys. Plasmas*, 1994, *1*, 1626.

[2.20] Monot, P.; et al. *Phys. Rev. Lett.* 1995, *74*, 2953.

[2.21] Borghesi, M.; et al. *Phys. Rev. Lett.* 1997, *78*, 879.

[2.22] Borisov, A.; et al. *Plasma Phys. Controlled Fusion*, 1995, *37*, 569.

[2.23] Malka, G.; et al. *Phys. Rev. Lett.* 1997, *78*, 3314.

[2.24] Malka, G.; et al. *Phys. Rev. Lett.* 1997, *79*, 2053.

[2.25] Marciak – Kozlowska, J.; Kozlowski, M. *Lasers in Engineering*, 2002, *12*, 17.

[2.26] Kozlowski, M.; Marciak – Kozlowska, J. *From Quarks to Bulk Matter*; Hadronic Press: Palm Harbor, FL, 2001.

[2.27] Synge, J. L. *The Relativistic Gas*; North-Holland Publisher Company, 1957.

[2.28] Ehrenfest, P. *Proc. Amsterdam Acad.* 1917, *20*, 200.

[2.29] Tangherlini, F. R. *Nuovo Cimento* 1963, *27*, 636.

[2.30] Whitrow, C. J. *Brit. J. Phil.* 1955, *6*, 13.

[2.31] Evans, L. C. *Partial Differential Equations*; American Mathematical Society, USA, 1998.

[2.32] Hashimoto, T.; et al. *Phys. Rev.* 1997, *B13*, 1119.

Chapter 3

SCHRÖDINGER – NEWTON HYPERBOLIC EQUATION FOR QUANTUM THERMAL PROCESSES

3.1. NON – LOCAL SCHRÖDINGER EQUATION

When M. Planck made the first quantum discovery he noted an interesting fact [3.1]. The speed of light, Newton's gravity constant and Planck's constant clearly reflect fundamental properties of the world. From them it is possible to derive the characteristic mass M_P, length L_P and time T_P with approximate values

$$L_P = 10^{-35}\ \text{m}$$
$$T_P = 10^{-43}\ \text{s}$$
$$M_P = 10^{-5}\ \text{g}\ .$$

Nowadays much of cosmology is concerned with "interface" of gravity and quantum mechanics.

After the *Alpha* moment – the spark in eternity [3.1] the space and time were created by "Intelligent Design" [3.2] at $t = T_P$. The enormous efforts of the physicists, mathematicians and philosophers investigate the *Alpha* moment. Scholars seriously discuss the Alpha moment – by all possible means: theological and physico-mathematical with growing complexity of theories. The most important result of these investigations is the anthropic principle and Intelligent Design theory (ID).

In this chapter we investigate the very simple question: how gravity can modify the quantum mechanics, i.e. the nonrelativistic Schrödinger equation (SE). We argue that SE with relaxation term describes properly the quantum behaviour of particle with mass $m < M_P$ and contains the part which can be interpreted as the pilot wave equation. For $m \to M_P$ the solution of the SE represent the strings with mass M_P.

The thermal history of the system (heated gas container, semiconductor or Universe) can be described by the generalized Fourier equation [3.3 – 3.5]

$$q(t) = -\int_{-\infty}^{t} \underbrace{K(t-t')}_{\text{thermal history}} \underbrace{\nabla T(t')dt'}_{\text{diffusion}}. \tag{3.1}$$

In Eq. (3.1) $q(t)$ is the density of the energy flux, T is the temperature of the system and $K(t-t')$ is the thermal memory of the system

$$K(t-t')=\frac{K}{\tau}\exp\left[-\frac{(t-t')}{\tau}\right], \tag{3.2}$$

where K is constant, and τ denotes the relaxation time.

As was shown in [3.3 – 3.5]

$$K(t-t')=\begin{cases} K\delta(t-t') & \text{diffusion} \\ K=\text{constant} & \text{wave} \\ \dfrac{K}{\tau}\exp\left[-\dfrac{(t-t')}{\tau}\right] & \text{damped wave or hyperbolic diffusion.} \end{cases}$$

The damped wave or hyperbolic diffusion equation can be written as:

$$\frac{\partial^2 T}{\partial t^2}+\frac{1}{\tau}\frac{\partial T}{\partial t}=\frac{D_T}{\tau}\nabla^2 T. \tag{3.3}$$

For $\tau\to 0$, Eq. (3.3) is the Fourier thermal equation

$$\frac{\partial T}{\partial t}=D_T\nabla^2 T \tag{3.4}$$

and D_T is the thermal diffusion coefficient. The systems with very short relaxation time have very short memory. On the other hand for $\tau\to\infty$ Eq. (3.3) has the form of the thermal wave (undamped) equation, or *ballistic* thermal equation. In the solid state physics the *ballistic* phonons or electrons are those for which $\tau\to\infty$. The experiments with *ballistic* phonons or electrons demonstrate the existence of the *wave motion* on the lattice scale or on the electron gas scale.

$$\frac{\partial^2 T}{\partial t^2}=\frac{D_T}{\tau}\nabla^2 T. \tag{3.5}$$

For the systems with very long memory Eq. (3.3) is time symmetric equation with no arrow of time, for the Eq. (3.5) does not change the shape when $t\to -t$.

In Eq. (3.3) we define:

$$\upsilon=\left(\frac{D_T}{\tau}\right), \tag{3.6}$$

velocity of thermal wave propagation and

$$\lambda = \upsilon\tau, \tag{3.7}$$

where λ is the mean free path of the heat carriers. With formula (3.6) equation (3.3) can be written as

$$\frac{1}{\upsilon^2}\frac{\partial^2 T}{\partial t^2} + \frac{1}{\tau\upsilon^2}\frac{\partial T}{\partial t} = \nabla^2 T. \tag{3.8}$$

From the mathematical point of view equation:

$$\frac{1}{\upsilon^2}\frac{\partial^2 T}{\partial t^2} + \frac{1}{D}\frac{\partial T}{\partial t} = \nabla^2 T$$

is the hyperbolic partial differential equation (PDE). On the other hand Fourier equation

$$\frac{1}{D}\frac{\partial T}{\partial t} = \nabla^2 T \tag{3.9}$$

and Schrödinger equation

$$i\hbar\frac{\partial \Psi}{\partial t} = -\frac{\hbar^2}{2m}\nabla^2\Psi \tag{3.10}$$

are the parabolic equations. Formally with substitutions

$$t \leftrightarrow it, \ \Psi \leftrightarrow T \tag{3.11}$$

Fourier equation (3.9) can be written as

$$i\hbar\frac{\partial \Psi}{\partial t} = -D\hbar\nabla^2\Psi \tag{3.12}$$

and by comparison with Schrödinger equation one obtains

$$D_T\hbar = \frac{\hbar^2}{2m} \tag{3.13}$$

and

$$D_T = \frac{\hbar}{2m}. \qquad (3.14)$$

Considering that $D_T = \tau v^2$ (3.6) we obtain from (3.14)

$$\tau = \frac{\hbar}{2mv_h^2}. \qquad (3.15)$$

Formula (3.15) describes the relaxation time for quantum thermal processes.
Starting with Schrödinger equation for particle with mass m in potential V:

$$i\hbar\frac{\partial\Psi}{\partial t} = -\frac{\hbar^2}{2m}\nabla^2\Psi + V\Psi \qquad (3.16)$$

and performing the substitution (3.11) one obtains

$$\hbar\frac{\partial T}{\partial t} = \frac{\hbar^2}{2m}\nabla^2 T - VT \qquad (3.17)$$

$$\frac{\partial T}{\partial t} = \frac{\hbar}{2m}\nabla^2 T - \frac{V}{\hbar}T. \qquad (3.18)$$

Equation (3.18) is Fourier equation (parabolic PDE) for $\tau = 0$. For $\tau \neq 0$ we obtain

$$\tau\frac{\partial^2 T}{\partial t^2} + \frac{\partial T}{\partial t} + \frac{V}{\hbar}T = \frac{\hbar}{2m}\nabla^2 T, \qquad (3.19)$$

$$\tau = \frac{\hbar}{2mv^2} \qquad (3.20)$$

or

$$\frac{1}{v^2}\frac{\partial^2 T}{\partial t^2} + \frac{2m}{\hbar}\frac{\partial T}{\partial t} + \frac{2Vm}{\hbar^2}T = \nabla^2 T.$$

with the substitution (3.11) equation (3.19) can be written as

$$i\hbar\frac{\partial\Psi}{\partial t} = V\Psi - \frac{\hbar^2}{2m}\nabla^2\Psi - \tau\hbar\frac{\partial^2\Psi}{\partial t^2}. \qquad (3.21)$$

The new term, relaxation term

$$\tau\hbar\frac{\partial^2\Psi}{\partial t^2} \tag{3.22}$$

describes the interaction of the particle with mass m with space-time. The relaxation time τ can be calculated as:

$$\tau^{-1} = \left(\tau_{e-p}^{-1} + ... + \tau_{Planck}^{-1}\right), \tag{3.23}$$

where, for example τ_{e-p} denotes the scattering of the particle m on the electron-positron pair ($\tau_{e-p} \sim 10^{-17}$ s) and the shortest relaxation time τ_{Planck} is the Planck time ($\tau_{Planck} \sim 10^{-43}$ s).

From equation (3.23) we conclude that $\tau \approx \tau_{Planck}$ and equation (3.21) can be written as

$$i\hbar\frac{\partial\Psi}{\partial t} = V\Psi - \frac{\hbar^2}{2m}\nabla^2\Psi - \tau_{Planck}\hbar\frac{\partial^2\Psi}{\partial t^2}, \tag{3.24}$$

where

$$\tau_{Planck} = \frac{1}{2}\left(\frac{\hbar G}{c^5}\right)^{\frac{1}{2}} = \frac{\hbar}{2M_p c^2}. \tag{3.25}$$

In formula (3.25) M_p is the mass Planck. Considering Eq. (3.25), Eq. (3.24) can be written as

$$i\hbar\frac{\partial\Psi}{\partial t} = -\frac{\hbar^2}{2m}\nabla^2\Psi + V\Psi - \frac{\hbar^2}{2M_p}\nabla^2\Psi + \frac{\hbar^2}{2M_p}\nabla^2\Psi - \frac{\hbar^2}{2M_p c^2}\frac{\partial^2\Psi}{\partial t^2}. \tag{3.26}$$

The last two terms in Eq. (3.26) can be defined as the *Bohmian* pilot wave

$$\frac{\hbar^2}{2M_p}\nabla^2\Psi - \frac{\hbar^2}{2M_p c^2}\frac{\partial^2\Psi}{\partial t^2} = 0, \tag{3.27}$$

i.e.

$$\nabla^2\Psi - \frac{1}{c^2}\frac{\partial^2\Psi}{\partial t^2} = 0. \tag{3.28}$$

It is interesting to observe that pilot wave Ψ does not depend on the mass of the particle. With postulate (3.28) we obtain from equation (3.26)

$$i\hbar\frac{\partial\Psi}{\partial t}=-\frac{\hbar^2}{2m}\nabla^2\Psi+V\Psi-\frac{\hbar^2}{2M_p}\nabla^2\Psi \tag{3.29}$$

and simultaneously

$$\frac{\hbar^2}{2M_p}\nabla^2\Psi-\frac{\hbar^2}{2M_pc^2}\frac{\partial^2\Psi}{\partial t^2}=0. \tag{3.30}$$

In the operator form Eq. (3.21) can be written as

$$\hat{E}=\frac{\hat{p}^2}{2m}+\frac{1}{2M_pc^2}\hat{E}^2, \tag{3.31}$$

where $\hat{E}$ and $\hat{p}$ denote the operators for energy and momentum of the particle with mass m. Equation (3.31) is the new dispersion relation for quantum particle with mass m. From Eq. (3.21) one can concludes that Schrödinger quantum mechanics is valid for particles with mass $m \ll M_P$. But pilot wave exists independent of the mass of the particles.

For particles with mass $m \ll M_P$ Eq. (3.29) has the form

$$i\hbar\frac{\partial\Psi}{\partial t}=-\frac{\hbar^2}{2m}\nabla^2\Psi+V\Psi. \tag{3.32}$$

In the case when $m \approx M_p$ Eq. (3.29) can be written as

$$i\hbar\frac{\partial\Psi}{\partial t}=-\frac{\hbar^2}{2M_p}\nabla^2\Psi+V\Psi, \tag{3.33}$$

but considering Eq. (3.30) one obtains

$$i\hbar\frac{\partial\Psi}{\partial t}=-\frac{\hbar^2}{2M_pc^2}\frac{\partial^2\Psi}{\partial t^2}+V\Psi \tag{3.34}$$

or

$$\frac{\hbar^2}{2M_pc^2}\frac{\partial^2\Psi}{\partial t^2}+i\hbar\frac{\partial\Psi}{\partial t}-V\Psi=0. \tag{3.35}$$

We look for the solution of Eq. (3.35) in the form

$$\Psi(x,t) = e^{i\omega t}u(x). \tag{3.36}$$

After substitution formula (3.36) to Eq. (3.35) we obtain

$$\frac{\hbar^2}{2M_p c^2}\omega^2 + \omega\hbar + V(x) = 0 \tag{3.37}$$

with the solution

$$\omega_1 = \frac{-M_p c^2 + M_p c^2\sqrt{1-\frac{2V}{M_p c^2}}}{\hbar}$$
$$\omega_2 = \frac{-M_p c^2 - M_p c^2\sqrt{1-\frac{2V}{M_p c^2}}}{\hbar} \tag{3.38}$$

for $\frac{M_p c^2}{2} > V$ and

$$\omega_1 = \frac{-M_p c^2 + iM_p c^2\sqrt{\frac{2V}{M_p c^2}-1}}{\hbar}$$
$$\omega_2 = \frac{-M_p c^2 - iM_p c^2\sqrt{\frac{2V}{M_p c^2}-1}}{\hbar} \tag{3.39}$$

for $\frac{M_p c^2}{2} < V$.

Both formulae (3.38) and (3.39) describe the string oscillation, formula (3.27) damped oscillation and formula (3.28) over damped string oscillation.

3.2. The Time Evolution of the Bohmian Pilot Wave

D. Bohm presented the pilot wave theory in 1952 and de Broglie had presented a similar theory in the mid 1920's. It was rejected in 1950's and the rejection had nothing to do with de Broglie and Bohm later works.

There is always the possibility that the pilot wave has a primitive, mind like property. That's how Bohm described it. We can say that all the particles in the Universe end even Universe have their own pilot waves, their own information. Then the consciousness for example is the very complicated receiver of the surrounding pilot wave fields.

In our paper [3.6], a study of the Newton-Schrödinger-Bohm (NSB) equation for the pilot wave was developed:

$$i\hbar\frac{\partial\Psi}{\partial t}=-\frac{\hbar^2}{2m}\nabla^2\Psi+V\Psi-\frac{\hbar^2}{2M_p}\nabla^2\Psi+\frac{\hbar^2}{2M_p}\left(\nabla^2\Psi-\frac{1}{c^2}\frac{\partial^2\Psi}{\partial t^2}\right). \tag{3.40}$$

In Eq. (3.40) m is the mass of the quantum particle and M_P is the Planck mass ($M_P \approx 10^{-5}$ g).

For elementary particles with mass $m << M_P$ we obtain from Eq. (3.40)

$$i\hbar\frac{\partial\Psi}{\partial t}=-\frac{\hbar^2}{2m}\nabla^2\Psi+V\Psi+\frac{\hbar^2}{2M_p}\left(\nabla^2\Psi-\frac{1}{c^2}\frac{\partial^2\Psi}{\partial t^2}\right) \tag{3.41}$$

and for macroscopic particles with $m >> M_P$ equation (3.40) has the form:

$$i\hbar\frac{\partial\Psi}{\partial t}=-\frac{\hbar^2}{2M_p}\nabla^2\Psi+\frac{\hbar^2}{2M_p}\left(\nabla^2\Psi-\frac{1}{c^2}\frac{\partial^2\Psi}{\partial t^2}\right)+V\Psi \tag{3.42}$$

or

$$i\hbar\frac{\partial\Psi}{\partial t}=-\frac{\hbar^2}{2M_pc^2}\frac{\partial^2\Psi}{\partial t^2}+V\Psi$$

and is dependent of m.

In the following we will discuss the pilot wave time evolution for the macroscopic particles, i.e. for particles with $m >> M_P$.

For V = const. we seek the solution of Eq. (3.42) in the form:

$$\Psi=e^{\gamma t}. \tag{3.43}$$

After substitution formula (3.43) to Eq. (3.42) one's obtains

$$M_P\gamma^2+\frac{2M_P^2c^2}{\hbar}\gamma-\frac{2M_P^2c^2}{\hbar^2}V=0 \tag{3.44}$$

with the solution

$$\gamma_{1,2} = -\frac{iM_Pc^2}{\hbar} \pm \frac{M_Pc^2}{\hbar}\sqrt{-1+\frac{2V}{M_Pc^2}}. \tag{3.45}$$

For a free particle, $V = 0$ we obtain:

$$\gamma_{1,2} = \begin{cases} 0, \\ -\dfrac{2M_Pc^2}{\hbar}i. \end{cases} \tag{3.46}$$

According to formulae (3.43) and (3.46) equation (3.42) has the solution

$$\Psi(t) = A + Be^{-\frac{2M_Pc^2 i}{\hbar}t}. \tag{3.47}$$

For $t = 0$ we put $\Psi(0) = 0$, then

$$\Psi(t) = A\left(1 - e^{-\frac{2it}{\tau_P}}\right),$$

where τ_P = Planck time

$$\tau_P = \frac{\hbar}{M_pc^2}. \tag{3.48}$$

From formula (3.48) we conclude that the free particle in reality is jittering with frequency $\omega = \tau^{-1}$ and quantum energy $E = \hbar\omega = 10^{19}$ GeV and period $T = 10^{-43}$ s.

3.3. Quantum Thermal Waves in Quantum Corrals

Recently has been a great interest in both ultrafast (femtosecond and attosecond) laser-induced kinetics and in nanoscale properties of matter. Particular attention has been attracted by phenomena that are simultaneously nanoscale and ultrafast [3.7 – 3.15]. Fundamentally nanosize eliminates effects of electromagnetic retardation and thus facilitates coherent ultrafast kinetics. On the applied side, nanoscale design of optoelectronic devices is justified if their operating times are ultrashort to allow for ultrafast computing and transmission of information.

One of the key problems of ultrafast/nanoscale physics is ultrafast excitation of nanosystem where the transferred energy localizes at a given site. Because the electromagnetic wavelength is on a much larger microscale it is impossible to employ light-wave focusing for that purpose. In paper [3.16] the method to use phase modulation of an exciting femtosecond pulse is proposed. This method of localization exist due to the fact that polar excitation (surface plasmous) in inhomogeneous nanosystems tend to be localized with their oscillation frequency (and, consequently, phase) correlated with position [3.17, 3.18].

The coherently controlled ultrafast energy localization in nanosystems introduced in paper [3.16], can have applications in different fields that require directed nanosize-selective excitation.

In recent years the advances in scanning tunneling microscopy (STM) made possible the manipulation of single atoms on top of a surface and the construction of quantum-nanometre scale structures of arbitrary shapes [3.19]. In particular, quantum corrals have been assembled by depositing a close line of atoms or molecules on Cu or noble metal surface [3.20 – 3.23]. These surfaces have the property that for small wave vectors parallel to the surface a parabolic band of two-dimensional (2D) surface states uncoupled to bulk states exists [3.24]. In quantum corrals the STM tip can exist standing wave pattern of the one electron de Broglie waves.

In this chapter we describe the thermal excitation of the de Broglie electron waves with attosecond laser pulses. With coherent control of the ultrashort laser pulses it is possible to concentrate the laser energy on the nanometer scale [3.16]. Following the results of paper [3.25] we will describe the temperature of the electron 2D gas with the help of the quantum hyperbolic heat transfer equation.

In the following we consider the 2D heat transfer phenomena described by the equation [3.25]:

$$\frac{1}{\upsilon^2}\frac{\partial^2 T}{\partial t^2}+\frac{1}{D}\frac{\partial T}{\partial t}+\frac{2Vm}{\hbar^2}T=\nabla^2 T, \tag{3.49}$$

where T is temperature of the 2D electron gas

$$T=T(x,y,t), \tag{3.50}$$

D is the thermal diffusion coefficient, V is the nonthermal potential and m is the mass of the heat carriers–electrons.

We seek solution of Eq. (3.49) in the form

$$T(x,y,t)=e^{-\frac{t}{2\tau}}u(x,y,t). \tag{3.51}$$

After substitution of Eq. (3.51) to Eq. (3.49) one obtains

$$\frac{1}{\upsilon^2}\frac{\partial^2 u}{\partial t^2}-\nabla^2 u+q^2 u=0, \tag{3.52}$$

Where

$$q^2 = \frac{2Vm}{\hbar^2} - \left(\frac{m\upsilon}{2\hbar}\right)^2 \tag{3.53}$$

for $D = \frac{\hbar}{m}$.

We can define the distortionless thermal wave as the wave which preserves the shape in the field of the potential V. The condition for conserving the shape can be formulated as

$$q^2 = \frac{2Vm}{\hbar^2} - \left(\frac{m\upsilon}{2\hbar}\right)^2 = 0. \tag{3.54}$$

When Eq. (3.54) holds Eq. (3.52) has the form

$$\frac{1}{\upsilon^2}\frac{\partial^2 u}{\partial t^2} - \nabla^2 u = 0 \tag{3.55}$$

and condition (3.54) can be written as

$$V\tau \sim \hbar, \tag{3.56}$$

where τ is the relaxation time

$$\tau = \frac{\hbar}{m\upsilon^2}. \tag{3.57}$$

We conclude that in the presence of the potential energy V one can observe the undisturbed thermal wave only when the Heisenberg uncertainty relaxation (3.56) is fulfilled.

In the subsequent we will consider the thermal relaxation of the 2D electron gas contained in 2D circular quantum corral with the radius r. In that case in polar coordinates equation (3.55) has the form

$$\frac{1}{r}\frac{\partial}{\partial r}\left(r\frac{\partial u}{\partial r}\right) + \frac{1}{r^2}\frac{\partial^2 u}{\partial \theta^2} = \frac{1}{\upsilon^2}\frac{\partial^2 u}{\partial t^2}, \tag{3.58}$$

where $0 < r < a$, $-\pi < \theta < \pi$ with boundary condition

$$\begin{aligned} &u(r,\theta,0) = f(r,\theta), \quad 0 < r < a, \quad -\pi < \theta \le \pi, \\ &\frac{\partial u}{\partial t}(r,\theta,0) = g(r,\theta), \quad 0 < r < a, \quad -\pi < \theta \le \pi. \end{aligned} \tag{3.59}$$

Using separation of variables $u(r,t) = R(r)T(t)$ yields the solution

$$
\begin{aligned}
u(r,\theta,t) = & \sum_n a_{on} J_0(\lambda_{on} r)\cos(\lambda_{on} vt) \\
& + \sum_{m,n} a_{mn} J_m(\lambda_{mn} r)\cos(m\theta)\cos(\lambda_{mn} ct) \\
& + \sum_{m,n} b_{mn} J_m(\lambda_{mn} r)\sin(m\theta)\cos(\lambda_{mn} ct) \\
& + \sum_n A_{on} J_0(\lambda_{on} r)\sin(\lambda_{on} ct) \\
& + \sum_{m,n} A_{mn} J_m(\lambda_{mn} r)\cos(m\theta)\sin(\lambda_{mn} ct) \\
& + \sum_{m,n} B_{mn} J_m(\lambda_{mn} r)\sin(m\theta)\sin(\lambda_{mn} ct),
\end{aligned}
\tag{3.60}
$$

where J_m represents the m-th Bessel function of the first kind, λ_{mn} represents the n-th zero of J_m, and the coefficients a_{on}, a_{mn}, b_{mn}, A_{on}, A_{mn} and B_{mn} can be find out in [3.26].

In the subsequent we present the numerical solution of the Eq. (3.55) for the thermal wave with velocity $\upsilon = 5 \cdot 10^{-3} c$, c = light velocity. Considering formula (3.57) for relaxation time we obtain

$$\tau = \frac{\hbar}{m\upsilon^2} = 160 \text{ as} \tag{3.61}$$

for $m = m_e$, $(= 0.51\,\text{MeV})$ i.e. for electrons and for mean free path of the electrons in the 2D electron gas:

$$\lambda_{mfp} = \upsilon\tau \approx 0.1\,\text{nm}. \tag{3.62}$$

From formula (3.62) we conclude that λ_{mfp} is of the order of the de Broglie'a wave length of the electron. It means that for 2D electron gas in quantum stadium the hyperbolic quantum thermal equation can be applied [3.27].

3.4. Klein – Gordon Thermal Equation with Casimir Potential for Attosecond Laser Pulse Interaction with Matter

The contemporary nanoelectronic develops the NEMS and MEMS structures in which the distance between the parts is of the order of nanometers. As was shown in monograph:

Form quarks to bulk matter the transport phenomena on the nanoscale depend on the second derivative in time. It is in contrast to the macroscale heat transport where the Fourier law (only the first derivative in time). The second derivative in time term describes the memory of the thermally excited medium. Considering the contemporary discussion of the role played by Casimir force in the NEMS and MEMS in this chapter we describe the heat signaling in the simple nanostructure the parallel plates heated by attosecond laser pulses. It will be shown that temperature field between plates depends on the distance of the plates (in nanoscale). As the result the attosecond laser pulse can be used as the tool for the investigation of Casimir effect on the performance of the NEMS and MEMS.

Vacuum energy is a consequence of the quantum nature of the electromagnetic field, which is composed of photons. A photon of frequency ω has energy $\hbar\omega$, where $\hbar$ is Planck constant. The quantum vacuum can be interpreted as the lowest energy state (or ground state) of the electromagnetic (EM) field that occurs when all charges and currents have been removed and the temperature has been reduced to absolute zero. In this state no ordinary photons are present. Nevertheless, because the electromagnetic field is a quantum system the energy of the ground state of the EM is not zero. Although the average value of the electric field $\langle E \rangle$ vanishes in ground state, the Root Mean Square of the field $\langle E^2 \rangle$ is not zero. Similarly the $\langle B^2 \rangle$ is not zero. Therefore the electromagnetic field energy $\langle E^2 \rangle + \langle B^2 \rangle$ is not equal zero. A detailed theoretical calculation tells that EM energy in each mode of oscillation with frequency ω is 0.5 $\hbar\omega$, which equals one half of amount energy that would be present if a single "real" photon of that mode were present. Adding up 0.5 $\hbar\omega$ for all possible modes of electromagnetic field gives a very large number for the vacuum energy E_0 in the quantum vacuum

$$E_0 = \sum_i \frac{1}{2}\hbar\omega_i. \tag{3.63}$$

The resulting vacuum energy E_0 is *infinity* unless a high frequency cut off is applied.

Inserting surfaces into the vacuum causes the modes of the EM to change. This change in the modes that are present occurs since the EM must meet the appropriate boundary conditions at each surface. Surface alters the modes of oscillation and therefore the surfaces alter the energy density corresponding to the lowest state of the EM field. In actual practice the change in E_0 is defined as follows

$$\Delta E_0 = E_0 - E_S \tag{3.64}$$

where E_0 is the energy in empty space and E_S is the energy in space with surfaces, i.e.

$$\Delta E_0 = \frac{1}{2}\sum_n^{\text{empty space}} \hbar\omega_n - \frac{1}{2}\sum_i^{\text{surface present}} \hbar\omega_i\,. \tag{3.65}$$

As an example let us consider a hollow conducting rectangular cavity with sides a_1, a_2, a_3. In that case for uncharged parallel plates with area A the attractive force between the plates is [3.28]:

$$F_{att} = -\frac{\pi^2 \hbar c}{240 d^4} A, \tag{3.66}$$

where d is the distance between plates. The force F_{att} is called the parallel plate Casimir force, which was measured in three different experiments [3.29 – 3.31].

Recent calculation show that for conductive rectangular cavities the vacuum forces on a given face can be repulsive (positive), attractive (negative) or zero depending on the ratio of the sides [3.32].

In paper [3.33] the first measurement of repulsive Casimir force was performed. For the distance (separation) $d \sim 0.1$ μm the repulsive force is of the order of 0.5 μN – for cavity geometry. In March 2001, scientist at Lucent Technology used attractive parallel plate Casimir force to actuate a MEMS torsion device [3.31]. Other MEMS (MicroElectroMechanical System) have been also proposed [3.34].

Standard Klein – Gordon equation reads:

$$\frac{1}{c^2}\frac{\partial^2 \Psi}{\partial t^2} - \frac{\partial^2 \Psi}{\partial x^2} + \frac{m^2 c^2}{\hbar^2}\Psi = 0. \tag{3.67}$$

In equation (3.67) Ψ is the relativistic wave function for particle with mass m, c is the light velocity and $\hbar$ is Planck constant. For massless particles $m = 0$ Eq. (3.67) is the Maxwell equation for photons. As was shown by Pauli and Weiskopf since relativistic quantum mechanical equation had to allow for creation and annihilation of particles, the Klein – Gordon describes spin – 0 bosons.

In monograph [3.27] the generalized Klein – Gordon thermal equation was developed

$$\frac{1}{\upsilon^2}\frac{\partial^2 T}{\partial t^2} - \nabla^2 T + \frac{m}{\hbar}\frac{\partial T}{\partial t} + \frac{2Vm}{\hbar^2} = 0. \tag{3.68}$$

In Eq. (3.68) T denotes temperature of the medium and v is the velocity of the temperature signal in the medium. When we extract the highly oscillating part of the temperature field,

$$T = e^{-\frac{t\omega}{2}} u(x,t), \tag{3.69}$$

where $\omega = \tau^{-1}$, and τ is the relaxation time, we obtain from Eq. (3.68) (1D case)

$$\frac{1}{\upsilon^2}\frac{\partial^2 u}{\partial t^2} - \frac{\partial^2 u}{\partial x^2} + q^2 u(x,t) = 0, \tag{3.70}$$

where

$$q^2 = \frac{2Vm}{\hbar^2} - \left(\frac{mv}{2\hbar}\right)^2. \tag{3.71}$$

When $q^2 > 0$ equation (3.70) is of the form of the Klein – Gordon equation in the potential field $V(x, t)$. For $q^2 < 0$ Eq. (3.70) is the modified Klein – Gordon equation. The discussion of the physical properties of the solution of equation (3.70) can be found in [3.27].

3.5. Attophysics of Thermal Phenomena in Carbon Nanotubes

The interaction of laser pulses with carbon nanotubes is a very interesting and new field of investigation. In nanotechnology the carbon nanotubes are the main parts of MEMS and in the future NEMS devices. In living organisms nanotubes build the skeleton of living cells. The exceptional properties of carbon nanotubes (CNTs), including ballistic transport and semiconducting behaviour with band-gaps in the range of 1 eV, have sparked a large number of theoretical [3.35 – 3.37] and experimental [3.38 – 3.40] studies. The possibility of using CNTs to replace crystalline silicon for high-performance transistors has resulted in an effort to reduce the size of CNT field-effect transistors (CNTFETs) in order to understand the scaling behaviour and the ultimate limit. In this context we discuss CNT transistors with channel lengths less than 20 nm but which have characteristics comparable to those of much larger silicon-based field effect transistors with similar channel lengths [3.41].

Since the first CNTFET was demonstrated in 1998 [3.42], their characteristics have been continuously and rapidly improved, particularly in the last few years [3.39, 3.40]. One critical aspect is the optimisation of the source and drain contacts to minimize the Schottky barrier (SB) due to the mismatch between the CNT and the contact metal work functions. As discussed by Guo et al. [3.35], the presence effective SBs of about 0.2 eV can severely affect the function of short-channel CNTFETs because electron injection through the SBs at higher drain-source biases reduces the on/off current ratio. In addition, substantial SBs reduce the on-current, as shown by the difference between Ti- and Pd-contacted CNTFETs [3.39, 3.40]. A further important aspect is the band gap of the CNTs, which can be expressed as 0.9/d eV, where d is the nanotube diameter in nanometers [3.43]. As shown by Javey et al.[11], it is possible to form virtually perfect contacts to 1.5-2 nm diameter nanotube CNTFETs using Pd contacts. These devices have on-conductivities close to quantum conduction limit $2G_0 = 155\ \mu s$. However, such CNTs have band gaps of only 0.4-0.5 eV, leading to high off-currents of short-channel devices [3.39]. Therefore, we have concentrated on CNTs with a smaller diameter about 0.7 – 1.1 nm which have a larger band gap about 0.8-1.3 eV and are suitable for short-channel devices [3.45]. Catalytic chemical vapor deposition (CCVD) enables CNTs with a narrow diameter distribution for the production of a significant number of device examples [3.45] or high current transistors [3.46] to be grown cleanly and selectively. In a recent publication it has been demonstrated that it is possible to grow small diameter CNTs for long channel CNTFETs gated using the substrate or e-beam defined top gates [3.47].

In this work we have investigated the thermal processes in carbon nanotubes with channel lengths below 20 nm. [3.48].

The breakthrough in the generation and detection of ultrashort, attosecond laser pulses with high harmonic generation technique [3.49, 3.50] has made this progress possible. This is the beginning of the attophysics age in which many-electron dynamics can be investigated in real time.

In new laser projects [3.51] the generation of a 100 GW-level attosecond X-ray pulses has been investigated. The relativistic multi-electron states can be generated by ultra-short (attosecond) high energy laser pulses.

The Dirac equation is used to describe the relativistic one electron state. In this chapter we develop and solve the Dirac type thermal equation for multi-electron states generated by the laser interaction with matter. The Dirac one dimensional thermal equation is applied to study of the generation of the positron-electron pairs. It is shown that the cross section is equal to the Thomson cross-section for electron-electron scattering.

As pointed in paper [3.52] a spin-flip occurs only when there is more than one dimension in space. Repeating the discussion for the derivation of the Dirac equation [3.52] for the case of one spatial dimension, it is found that the Dirac matrices and can be reduced to 2x2 matrices which can be represented by the Pauli matrices [3.52]. This fact simply implies that if there is only one spatial dimension, there is no spin. It is instructive to show explicitly how to derive the 1+1 dimensional Dirac equation.

As discussed in textbooks [3.52, 3.53] a wave equation which satisfies relativistic covariance in space-time as well as the probabilistic interpretation should have the form:

$$i\hbar\frac{\partial}{\partial t}\Psi(x,t)=\left[c\alpha\left(-i\hbar\frac{\partial}{\partial x}\right)+\beta m_0c^2\right]\Psi(x,t). \tag{3.72}$$

To obtain the relativistic energy-momentum relation $E^2=(pc)^2+m_0^2c^4$ we postulate that (3.72) coincides with the Klein-Gordon equation

$$\left[\frac{\partial^2}{\partial(ct)^2}-\frac{\partial^2}{\partial x^2}+\left(\frac{m_0c}{\hbar}\right)^2\right]\Psi(x,t)=0. \tag{3.73}$$

By comparing (3.72) and (3.73) it is easily seen that $\boldsymbol{\alpha}$ and $\boldsymbol{\beta}$ must satisfy

$$\boldsymbol{\alpha}^2-\boldsymbol{\beta}^2=1, \qquad \boldsymbol{\alpha\beta}+\boldsymbol{\beta\alpha}=0. \tag{3.74}$$

Any two of the Pauli matrices can satisfy these relations. Therefore, we may choose $\boldsymbol{\alpha}=\boldsymbol{\sigma}_x$ and $\boldsymbol{\beta}=\boldsymbol{\sigma}_z$ and we obtain:

$$i\hbar\frac{\partial}{\partial t}\Psi(x,t)=\left[c\sigma_x\left(-i\hbar\frac{\partial}{\partial x}\right)+\sigma_z m_0 c^2\right]\Psi(x,t), \tag{3.75}$$

where $\Psi(x,t)$ is a 2-component spinor.

The Eq. (3.75) is the Weyl representation of the Dirac equation. We perform a phase transformation on $\Psi(x,t)$ letting $u(x,t)=\exp\left(\frac{imc^2t}{\hbar}\right)\Psi(x,t)$. Calling u's upper (respectively, lower) component $u_+(x,t)$, $u_-(x,t)$; it follows from (3.75) that $u_\pm$ satisfies

$$\frac{\partial u_\pm(x,t)}{\partial t}=\pm c\frac{\partial u_\pm}{\partial x}+\frac{im_0c^2}{\hbar}(u_\pm-u_\mp). \tag{3.76}$$

Following the physical interpretation of the equation (3.76) it describes the relativistic particle (mass m_0) propagating at the speed of light c with a certain *chirality* (like a two component neutrino) except that at random times it flips in both the direction of propagation (by 180°) and chirality.

In monograph [3.54] we considered a particle moving on the line with a fixed speed w and supposed that from time to time it suffers a complete reversal of direction, $u(x,t)\Leftrightarrow v(x,t)$, where $u(x,t)$ denotes the expected density of particles at x and at time t moving to the right, and $v(x, t)$ = expected density of particles at x and at time t moving to the left. In the following we change the abbreviation

$$\begin{aligned} u(x,t)&\to u_+,\\ v(x,t)&\to u_-. \end{aligned} \tag{3.77}$$

Following the results of the paper [3.40] we obtain for the $u_\pm(x,t)$ the following equations

$$\begin{aligned} \frac{\partial u_+}{\partial t}&=-w\frac{\partial u_+}{\partial x}-\frac{w}{\lambda}((1-k)u_+-ku_-),\\ \frac{\partial u_-}{\partial t}&=w\frac{\partial u_-}{\partial x}+\frac{w}{\lambda}(ku_++(k-1)u_-). \end{aligned} \tag{3.78}$$

In equation (3.78) $k(x)$ denotes the number of the particles which are moving in left (right) direction after the scattering at x. The mean free path for scattering is equal λ, $\lambda=w\tau$, where τ is the relaxation time for scattering.

Comparing equations (3.76) and (3.78) we conclude that the form of both equations is the same. In the subsequent we will call the set of the equations (3.78) *the Dirac equation* for the particles with velocity w, mean free path λ.

For thermal processes we define $T_{+,-}$ ≡ the temperature of the particles with chirality + and – respectively and with analogy to equation (3.78) we obtain:

$$\begin{aligned}\frac{\partial T_+}{\partial t} &= -w\frac{\partial T_+}{\partial x} - \frac{w}{\lambda}\left((1-k)T_+ - kT_-\right),\\ \frac{\partial T_-}{\partial t} &= w\frac{\partial T_-}{\partial x} + \frac{w}{\lambda}\left(kT_+ + (k-1)T_-\right),\end{aligned} \tag{3.79}$$

where $\frac{w}{\lambda} = \frac{1}{\tau}$.

In one dimensional case we introduce one dimensional cross section for scattering

$$\sigma(x,t) = \frac{1}{\lambda(x,t)}. \tag{3.80}$$

In the stationary state thermal transport phenomena $\frac{\partial T_{+,-}}{\partial t} = 0$ and Eq. (3.79) can be written as

$$\begin{aligned}\frac{dT_+}{dx} &= -\sigma\left((1-k)T_+ + kT_-\right),\\ \frac{dT_-}{dx} &= \sigma(k-1)T_- + \sigma k T_+.\end{aligned} \tag{3.81}$$

After the differentiation of the equation (3.81) we obtain for $T_+(x)$

$$\frac{d^2T_+}{dx} - \frac{1}{\sigma k}\frac{d}{dx}(\sigma k)\frac{dT_+}{dx} + T_+\left[\sigma^2(2k-1) + \frac{d\sigma}{dx}(1-k) + \frac{\sigma(k-1)}{\sigma k}\frac{d(\sigma k)}{dx}\right] = 0. \tag{3.82}$$

Equation (3.82) can be written in a compact form

$$\frac{d^2T_+}{dx^2} + f(x)\frac{dT_+}{dx} + g(x)T_+ = 0, \tag{3.83}$$

where

$$f(x) = -\frac{1}{\sigma}\left(\frac{\sigma}{k}\frac{dk}{dx} + \frac{d\sigma}{dx}\right),$$

$$g(x) = \sigma^2(x)(2k-1) - \frac{\sigma}{k}\frac{dk}{dx}.$$

In the case for constant $\frac{dk}{dx} = 0$ we obtain

$$f(x) = -\frac{1}{\sigma}\frac{d\sigma}{dx},$$
$$g(x) = \sigma^2(x)(2k-1). \tag{3.84}$$

With functions $f(x)$, $g(x)$ described by formula (3.84) the general solution of Eq. (3.83) has the form:

$$T_+(x) = C_1 e^{(1-2k)^{\frac{1}{2}}\int \sigma(x)dx} + C_2 e^{-(1-2k)^{\frac{1}{2}}\int \sigma(x)dx} \tag{3.85}$$

and

$$T_-(x) = \frac{\left[(1-k)+(1-2k)^{\frac{1}{2}}\right]}{k} \times \left[C_1 e^{(1-2k)^{\frac{1}{2}}\int \sigma(x)dx} + \frac{(1-k)-(1-2k)^{\frac{1}{2}}}{(1-k)+(1-2k)^{\frac{1}{2}}} C_2 e^{-(1-2k)^{\frac{1}{2}}\int \sigma(x)dx}\right]. \tag{3.86}$$

The equations (3.85) and (3.86) describe three different modes for heat transport. For $k = \frac{1}{2}$ we obtain $T_+(x) = T_-(x)$ while for $k > \frac{1}{2}$, i.e. for heat carrier generation $T_+(x)$ and $T_-(x)$ oscillate for $(1-2k)^{\frac{1}{2}}$ is a complex number. For $k < \frac{1}{2}$ i.e. for absorption $T_+(x)$ and $T_-(x)$ decrease as the function of x.

In the following we shall consider the solution of Eq. (3.80) for Cauchy conditions:

$$T_+(0) = T_0, \quad T_-(a) = 0. \tag{3.87}$$

Boundary conditions (3.87) describes the generation of heat carriers by illuminating the left end of one dimensional slab (with length a) by laser pulse. From equations (3.85) and (3.86) we obtain:

$$T_+(x)=\frac{2T_0e^{[f(0)-f(a)]}}{1+\beta e^{2[f(0)-f(a)]}}\times\frac{(1-2k)^{\frac{1}{2}}\cosh[f(x)-f(a)]+(k-1)\sinh[f(x)-f(a)]}{(1-2k)^{\frac{1}{2}}-(k-1)}, \quad (3.88)$$

$$T_-(x)=\frac{2T_0e^{2[f(0)-f(a)]}\left[(k-1)+(1-2k)^{\frac{1}{2}}\right]\sinh[f(x)-f(a)]}{\left(1+\beta e^{-2[f(a)-f(0)]}\right)k}. \quad (3.89)$$

In equation (3.88) and (3.89)

$$\beta=\frac{(1-2k)^{\frac{1}{2}}+(k-1)}{(1-2k)^{\frac{1}{2}}-(k-1)} \quad (3.90)$$

and

$$\begin{aligned} f(x)&=(1-2k)^{\frac{1}{2}}\int\sigma(x)dx,\\ f(0)&=(1-2k)^{\frac{1}{2}}\left[\int\sigma(x)dx\right]_0,\\ f(a)&=(1-2k)^{\frac{1}{2}}\left[\int\sigma(x)dx\right]_a. \end{aligned} \quad (3.91)$$

Using equations (3.88) and (3.89) for $T_+(x)$ and $T_-(x)$ we define the asymmetry $A(x)$ of the temperature $T(x)$

$$A(x)=\frac{T_+(x)-T_-(x)}{T_+(x)+T_-(x)} \quad (3.92)$$

$$A(x)=\frac{\frac{(1-2k)^{\frac{1}{2}}}{(1-2k)^{\frac{1}{2}}-(k-1)}\cosh[f(x)-f(a)]-\frac{1-2k}{(1-2k)^{\frac{1}{2}}-(k-1)}\sinh[f(x)-f(a)]}{\frac{(1-2k)^{\frac{1}{2}}}{(1-2k)^{\frac{1}{2}}-(k-1)}\cosh[f(x)-f(a)]-\frac{1}{(1-2k)^{\frac{1}{2}}-(k-1)}\sinh[f(x)-f(a)]} \quad (3.93)$$

Based on equation (3.93) we conclude that for elastic scattering, i.e. when $k=\frac{1}{2}$, $A(x)=0$, and for $k\neq\frac{1}{2}$, $A(x)\neq 0$.

In the monograph [3.53] we introduced the relaxation time τ for quantum heat transport

$$\tau = \frac{\hbar}{m\upsilon^2}. \tag{3.94}$$

In equation (3.94) m denotes the mass of heat carriers electrons and $\upsilon = \alpha c$, where α is the fine structure constant for electromagnetic interactions. As was shown in monograph [3.54], τ is also the lifetime for positron-electron pairs in vacuum. When the duration of the laser pulse is less than τ, the hyperbolic transport equation must be used to describe transport phenomena. Recently, the structure of water was investigated using attosecond (10^{-18}s) resolution [3.55]. Considering that $\tau \approx 10^{-17}$ s we argue that to study performed in [3.55] open the new field for investigation of laser pulse with matter. In order to apply the equations (3.80) to attosecond laser induced phenomena we must know the cross section $\sigma(x)$. Considering equations (3.80) and (3.94) we obtain

$$\sigma(x) = \frac{m\upsilon}{\hbar} = \frac{me^2}{\hbar^2} \tag{3.95}$$

and it so happens that $\sigma(x)$ is the Thomson cross section for electron-electron scattering.

Based on equation (3.95) the solution of Cauchy problem has the form:

$$T_+(x) = \frac{2T_0 e^{-(1-2k)^{\frac{1}{2}}\frac{me^2}{\hbar^2}a}}{\left[1+\beta e^{-2(1-2k)^{\frac{1}{2}}\frac{me^2}{\hbar^2}a}\right]} \times$$

$$\frac{(1-2k)^{\frac{1}{2}}\cosh\left[(1-2k)^{\frac{1}{2}}\frac{me^2}{\hbar^2}(x-a)\right] + (k-1)\sinh\left[(1-2k)^{\frac{1}{2}}\frac{me^2}{\hbar^2}(x-a)\right]}{(1-2k)^{\frac{1}{2}} - (k-1)},$$

$$T_-(x) = \frac{2T_0 e^{-\frac{(1-2k)^{\frac{1}{2}} me^2 a}{\hbar^2}}\left[(k-1)-(1-2k)^{\frac{1}{2}}\right]}{\left(1+\beta e^{-2(1-2k)^{\frac{1}{2}}\frac{me^2}{\hbar^2}a}\right)k} \times$$

$$\sinh\left[(1-2k)^{\frac{1}{2}}\frac{me^2}{\hbar^2}(x-a)\right]. \tag{3.96}$$

REFERENCES

[3.1] Barbour, J. *The End of Time*; Oxford University Press, 2000.
[3.2]. Addicot, J. F. *Ohio State Law Journal*, 2003, *64*, 125.
[3.3]. Kozlowski, M.; Marciak – Kozlowska, J. *Foundations of Physics Letters*, 1997, *10*, 295.
[3.4]. Kozlowski, M.; Marciak – Kozlowska, J. *Foundations of Physics Letters*, 1997, *10*, 599.
[3.5]. Kozlowski, M.; Marciak – Kozlowska, J. *Foundations of Physics Letters*, 1999, *12*, 93.
[3.6]. Kozlowski, M.; Marciak – Kozlowska, J. (2004). Schrödinger – Newton wave mechanics. The model. arXiv:0402069.
[3.7]. Link, S.; et al. *Phys. Rev.* 2000, *B61*, 6086.
[3.8]. Lamprecht, B.; et al. *Phys. Rev. Lett.* 2000, *84*, 4721.
[3.9]. Klar, T.; et al. *Phys. Rev. Lett.* 1998, *80*, 4249.
[3.10]Lamprecht B.; et al. *Phys. Rev. Lett.* 1999, *83*, 4421.
[3.11]Klein-Wiele, J. H.; et al. *Phys. Rev. Lett.* 1998, *80*, 45.
[3.12]Stietz, F.; et al. *Phys. Rev. Lett.* 2000, *84*, 5149.
[3.13]Kulcsar G.; et al. *Phys. Rev. Lett.* 2000, *84*, 5149.
[3.14]Perner, M.; et al. *Phys. Rev. Lett.* 2000, *85*, 792.
[3.15]Stockman, M. J.; et al. *Phys. Rev.* 2000, *62*, 10494.
[3.16]Stockman, M. J.; et al. *Phys. Rev. Lett.* 2002, *88*, 067402-I.
[3.17]Stockman, M. J.; et al. *Phys. Rev. Lett.* 1997, *79*, 4562.
[3.18]Stockman, M. J.; et al. *Phys. Rev.* 1996, *B53*, 2183.
[3.19]Eigler, D. M.; Schweizer, E. K. *Nature* 1990, *344*, 524.
[3.20]Crommic, M. F.; et al. *Science* 1993, *262*, 218.
[3.21]Heller, E. J.; et al. *Nature* 1994, *369*, 464.
[3.22]Manoharan, H. C.; et al. *Nature* 2000, *403*, 512.
[3.23]Manoharan, H. C. PASI Conference Physics and Technology at the Nanometer Scale (Costa Rica, June-July 2001).
[3.24]Hulbert, S. L.; et al. *Phys. Rev.* 1985, *B31*, 6815.
[3.25]Marciak – Kozlowska, J.; Kozlowski, M. (2004). The thermal wave induced by ultra – short laser pulses in n – dimensional space – time. http://lanl.arxiv.org/cond-mat/0402159.
[3.26]Abbel, M. L.; Braselton, J. P. *Mathematica by Example*; A.P. Professional, Boston, MA, 1994.
[3.27]Kozlowski, M.; Marciak – Kozlowska, J. *From Quarks to Bulk Matter*; Hadronic Press: Palm Harbor, FL, 2001.
[3.28]Brown, L.; Maclay, J. *Phys. Rev.* 1969, *184*, 1272.
[3.29]Lamoroux, S. *Phys. Rev. Lett.* 1997, *78*, 5.
[3.30]Mohidem, U.; et al. *Phys. Rev. Lett.* 1998, *81*, 4549.
[3.31]Chan, H. B.; et al. *Science* 2001, *291*, 1941.
[3.32]Maclay, J. *Phys. Rev.* 2000, *A61*, 052110.
[3.33]Maclay, J.; et al. *Phys. Rev.* 2000, *A61*, 052110.
[3.34]Serry, M.; et al. *J. Microelectromechanical System* 1995, *4*, 193.
[3.35]Guo, J.; et al. *M. IEEE Trans. Nanotechnol.* 2003, *2,* 329-334.

[3.36]Guo, J.; et al. (2003). Predicted performance advantages of carbon nanotube transistors. arXiv :03 09039.
[3.37]Castro, A. H.; et al. *Proc. SPIE* 2004, *5276,* 1-10.
[3.38]Graham, A. P.; et al. *Relat. Mater.* 2004, *13,* 1296-1300.
[3.39]Javey, A.; et al. *Appl. Phys. Lett.* 2002, *80,* 3817-38 19.
[3.40]Wind, S. J.; et al. *Phys. Rev. Lett.* 2002, *80*, 3817.
[3.41]Yu, B.; et al. *IEEE International Electron Devices Meeting Technical Digest* 2002, *25,* 1-254.
[3.42]Tans, S. J.; Verschueren, A. R. M.; Dekker, C. *Nature* 1998, *393,* 49-52.
[3.43]McEuen, P.; et al. *IEEE Trans. Nanotechnol.* 2002, *1,* 78-85.
[3.44]Javey, A.; et al. *H. Nature* 2003, *424,* 654-657.
[3.45]Kreupl, F.; et al. (2004). Carbon nanotubes in microelectronic applications. arXiv:0410360.
[3.46]Seidel, R. V.; et al. *Nano Letters* 2004, *4*, 831.
[3.47]Graham, A. P.; et al. *Nano Letters* 2006, *6*, 2660.
[3.48]Seidel, R. V.; et al. *Nano Letters* 2005, *5*, 147.
[3.49]Dresher, M.; et al. *Science* 2001, *291*, 1923.
[3.50]Nikura, H.; et al. *Nature* 2003, *421*, 826.
[3.51]Saldin, E. L.; et al. (2004). A new technique to genetare 100 GW – level attosecond X-ray pulses from the X-ray SASE-FEL. http://lanl.arxiv.org/physics/0403067.
[3.52]Greiner, W. *Relativistic Quantum Mechanics*; Springer Verlag, Berlin, 1990.
[3.53]Bjorken, J. D.; Drell, S. D. *Relativistic Quantum Mechanics*; McGraw Hill: New York, NY, 1964.
[3.54]Kozlowski, M.; Marciak – Kozlowska, J. *Thermal Processes Using Attosecond Laser Pulses*; Optical Science 121; Springer: New York, NY, 2006.
[3.55]Abbamonte, P.; et al. *Phys. Rev. Lett.* 2004, *92*, 23740 1-1.

Chapter 4

PROCA EQUATION FOR RELATIVISTIC THERMAL PROCESSES

4.1. TRANSPORT PHENOMENA IN MINKOWSKI SPACETIME

The relativistic formulation of thermodynamics was taken up already by Einstein himself and by several other physicists, notably Planck and von Laue. The principal result at this stage was the following

$$T = T^{O}\sqrt{1-\tfrac{\upsilon^2}{c^2}} \tag{4.1}$$

In this equation T is the absolute temperature in a system which is moving with a velocity υ with respect to the rest system (which is indicated by a superscript *0*)

In the paper by Ott [4.1] the traditional formulation (4.1) was questioned. Instead of the transformation (4.1) for the temperature T, Ott requires

$$T = \frac{T^{O}}{\sqrt{1-\tfrac{\upsilon^2}{c^2}}} \tag{4.2}$$

In this chapter considering hyperbolic heat transport equation the temperature transformation equation will be obtained. It will be shown that the Ott formulation is valid Lorentz transformation for temperatures.

As was shown in monograph [4.2] the master equation for the heat transport induced by ultra-short laser pulses can be written as:

$$\vec{q} + \tau\frac{\partial\vec{q}}{\partial t} = -\kappa\frac{\partial T}{\partial x} \tag{4.3}$$

$$\frac{\partial\vec{q}}{\partial x} + c_V\frac{\partial T}{\partial t} = 0 \tag{4.4}$$

In Eq. (4.3) $\vec{q}$ is the heat current, T denotes temperature, τ is the relaxation time and κ is the heat conduction coefficient. In Eq. (4.4) c_V is the specific heat at constant volume. For quantum limit of the heat transport τ is equal

$$\tau = \frac{\hbar}{m\upsilon^2}, \tag{4.5}$$

where m is the mass of heat carriers and $\upsilon = \alpha c$. The constant α_i is the coupling constant, $\alpha_1 = 1/137$ for electromagnetic interaction and $\alpha_2 = 0.15$ for strong interaction, c is the vacuum light speed.

From Eqs (4.1) and (4.2) hyperbolic diffusion equation for temperature can be derived

$$\tau\frac{\partial^2 T}{\partial t^2} + \frac{\partial T}{\partial t} = D\frac{\partial^2 T}{\partial x^2}, \tag{4.6}$$

where thermal diffusion coefficient, D,

$$D = \frac{\kappa}{c_V}.$$

The basic principle of the special relativity theory can be stated as:

All physics laws look the same in all inertial references frames. Equation (4.4) is the conservation of thermal energy. Let us consider two infinite rods K and K', where K' is moving with velocity υ parallel to axis X. From the relativity principle we obtain

$$\begin{aligned} &\frac{\partial \vec{q}}{\partial x} + c_V\frac{\partial T}{\partial t} = 0 \text{ in frame } K \\ &\frac{\partial \vec{q}\,'}{\partial x'} + c_V\frac{\partial T'}{\partial t'} = 0 \text{ in frame } K' \end{aligned} \tag{4.7}$$

As can be easily shown the $\vec{q}$, $c_V T$ and $\vec{q}\,'$, $c_V T'$ are transformed according to Lorentz transformation:

$$\begin{aligned} &q'_{x'} = \gamma(q_x - \upsilon c_V T) \\ &q'_{y'} = q_y \\ &q'_{z'} = q_z \\ &c_V T' = \gamma\left[c_V T - \frac{\upsilon}{c^2}q_x\right] \end{aligned} \tag{4.8}$$

And

$$
\begin{aligned}
q_x &= \gamma(q'_{x'} + \upsilon c_V T') \\
q_y &= q'_{y'} \\
q_z &= q'_{z'} \\
c_V T &= \gamma\left[c_V T' + \frac{\upsilon}{c^2} q'_x\right]
\end{aligned}
\tag{4.9}
$$

The current q_x and temperature T form the four vectors.
The space time interval Δ,

$$
\Delta = q_x^2 - c^2 c_V T^2 = q'^2_{x'} - c^2 c_V T'^2 \tag{4.10}
$$

is invariant under the Lorentz transformations (4.8) and (4.9).

Let us consider the case where there is no heat current in rod K, i.e. when the rod has the temperature T = const and $\nabla T = 0$. In that case $q_x = 0$ and from formulae (4.8) we obtain

$$
\begin{aligned}
&q'_{x'} = -\gamma\ \upsilon\ c_V T \\
&c_V T' = \gamma\ c_V T; \qquad T' = \gamma\ T \\
&\quad T' > T \quad \text{for} \quad \gamma > 1
\end{aligned}
\tag{4.11}
$$

Formula (4.11) is in agreement with Ott result. As the result we obtain that temperatures of the rods are different. And the moving rod observes the greater temperature $T' = \gamma T$.

From Eq. (4.10) we conclude that

$$
q'^2_{x'} = c^2 c_V \left(T'^2 - T^2\right) \tag{4.12}
$$

and $q_x^2 > 0,\ q'^2_{x'} > 0$. It means that the heat current is directed parallel to the moving of the rod K in the reference frame K'.

4.2. On the Possible Thermal Tachyons

The square of the neutrino mass was measured in tritium beta decay experiments by fitting the shape of the beta spectrum near endpoint. In many experiments it has been found to be negative. According to the results of paper [4.3]

$$
m^2(\upsilon_e) = -2.5\ \text{eV}^2.
$$

Based on special relativity superluminal particles, i.e. particles with $m^2 < 0$ and $\upsilon > c$ were proposed and discussed in papers [4.3, 4.4]. In this chapter we investigate the possibility of the existence of the superluminal particles from the point of view the hyperbolic heat transport equation. It will be shown that hyperbolic heat transport equation is invariant under transformation $c^2 \to -c^2$, i.e. for transformation $c \to ic$. The new Lorentz transformation for $-c^2$ and formula for kinetic energy will be developed. It will be shown that for $\frac{E_k}{mc^2} > 1$ the speed of particles is decreasing for increased kinetic energy E_k.

In our monograph the hyperbolic Heaviside transport equation for attosecond laser pulses was obtained [4.2]

$$\frac{1}{\upsilon^2}\frac{\partial^2 T}{\partial t^2} + \frac{1}{D_T}\frac{\partial T}{\partial t} = \nabla^2 T. \tag{4.13}$$

In this equation T is the absolute temperature, υ denotes the speed of the thermal disturbance, and D_T is diffusion coefficient for thermal phenomena. Equation (4.13) describes the damped thermal wave propagation. Recently, the observation of the thermal wave in GaAs films exposed to ultra-short laser pulses was presented [4.4]. In the subsequent we will consider the mathematical structure of the Eq. (4.13) which can be interested for both the experimentalist as well as the theorist involved in ultrahigh ultra-short laser pulses. In the monograph [4.2] it was shown that speed υ, and diffusion coefficient D_T in Eq. (4.13) can be written as

$$\upsilon = \alpha c, \qquad D = \frac{\hbar}{m}. \tag{4.14}$$

In formula (4.14) α is the electromagnetic fine structure constant, $\hbar$ is the Planck constant, m is the mass of the heat carrier, and c is light velocity.

Let us consider the transformation

$$c \to ic \tag{4.15}$$

for the Eq. (4.13). First of all we note that

$$\upsilon'^2 \to \alpha^2 c^2 = \upsilon^2, \qquad D'_T = D_T. \tag{4.16}$$

We conclude that the Eq. (4.13) is invariant under the transformation (4.15). One can say that Eq. (4.13) is valid for the universe for which $-c^2$ is the invariant constant.

In our Universe the Lorentz transformation has the form

$$x' = \frac{x - \upsilon t}{\sqrt{1 - \frac{\upsilon^2}{c^2}}},$$

$$t' = \frac{t - \frac{\upsilon}{c^2} x}{\sqrt{1 - \frac{\upsilon^2}{c^2}}}. \tag{4.17}$$

With the transformation (4.15) we obtain from formula (4.17)

$$x' = \frac{x - \upsilon t}{\sqrt{1 + \frac{\upsilon^2}{c^2}}},$$

$$t' = \frac{t + \frac{\upsilon}{c^2} x}{\sqrt{1 - \frac{\upsilon^2}{c^2}}}. \tag{4.18}$$

For the space-time interval we obtain

$$x^2 + c^2 t^2 = x'^2 + c^2 t'^2$$

as in c^2 special relativity.

For $-c^2$ SR we have new formula for the velocities

$$\upsilon' = \frac{\upsilon - V}{1 + \frac{\upsilon V}{c^2}}. \tag{4.19}$$

For speed $\upsilon = ic$, from formula (4.19) we obtain $\upsilon' = ic$, i.e. object with speed ic has the same speed in all inertial reference frames. Now, we consider the formulae for total energy and momentum of the particle with mass m,

$$E = \frac{-mc^2}{\sqrt{1 + \frac{\upsilon^2}{c^2}}}, \qquad p = \frac{m\upsilon}{\sqrt{1 + \frac{\upsilon^2}{c^2}}}. \tag{4.20}$$

From formula (4.20) we obtain

$$\frac{E}{p} = -\frac{c^2}{\upsilon}. \tag{4.21}$$

For objects with speed $\upsilon = ic$ we obtain from formula (4.21)

$$\frac{E}{p} = ic. \tag{4.22}$$

Considering formula

$$E = \sqrt{-p^2c^2 + m^2c^4} \tag{4.23}$$

we obtain, for $m = 0$

$$E = ipc, \qquad \frac{E}{p} = ic. \tag{4.24}$$

We conclude that objects with masses $m = 0$, have speed $\upsilon = ic$ in universe. We calculate the kinetic energy, E_k

$$E_k = -mc^2(\gamma - 1), \qquad \gamma = \frac{1}{\sqrt{1 + \frac{\upsilon^2}{c^2}}}. \tag{4.25}$$

From formula (4.25) we deduce the ratio $\frac{\upsilon^2}{c^2}$.

$$\frac{\upsilon^2}{c^2} = \frac{1 - \left(1 - \frac{E_k}{mc^2}\right)^2}{\left(1 - \frac{E_k}{mc^2}\right)^2}. \tag{4.26}$$

It is quite interesting that in $-c^2$ universe $\frac{\upsilon^2}{c^2}$ is singular for $E_k = mc^2$, i.e. when kinetic energy of the objects equals its internal energy. For the c^2 universe the formula which describes $\frac{\upsilon^2}{c^2}$ reads

$$\frac{\upsilon^2}{c^2} = \frac{\left(1+\dfrac{E_k}{mc^2}\right)^2 - 1}{\left(1+\dfrac{E_k}{mc^2}\right)^2}. \tag{4.27}$$

In monograph [4.2] we show that the Eq. (4.13) describes the propagation of heatons, quanta of thermal field $T(x,t)$. For quantum heat transfer equation (4.13) we seek solution in the form

$$T(x,t) = e^{-\frac{t}{2\tau}} u(x,t). \tag{4.28}$$

After substitution of Eq. (4.28) into Eq. (4.13) one obtains

$$\frac{1}{\upsilon^2}\frac{\partial^2 u}{\partial t^2} - \frac{\partial^2 u}{\partial x^2} + q^2 u(x,t) = 0, \tag{4.29}$$

where

$$q^2 = -\left(\frac{m\upsilon}{2\hbar}\right)^2. \tag{4.30}$$

It is interesting to observe that if we introduce the imaginary mass $m^* = im$ then formula can be written as

$$q^2 = -\left(\frac{m^*\upsilon}{2\hbar}\right)^2, \tag{4.31}$$

and Eq. (4.29) is the Klein-Gordon for *heatons* with imaginary mass $m^* = im$. According to the results of our monograph [4.2] we can call *heatons* with imaginary mass m^* the *tachyons*, i.e. particles with $\upsilon > c$.

One can conclude that the irradiation of the matter with attosecond laser pulse can produce the thermal *tachyons*, which propagate with speed $\upsilon > c$. It must be stressed that the fact that $\upsilon > c$ does not violate the special relativity as described in the beginning. Moreover, the *thermal tachyons* fulfils the same thermal transport equation (4.13) as the particles with $\upsilon < c$. The experimental method best suited for the observation of the thermal *heatons* is the TOF (Time of Flight) measurement of the velocity of the emitted particles.

On the experimental ground the existence of the *tachyons* is still an open question. However, the results of papers [4.3, 4.4] strongly suggest the existence of the particles with masses $m^* = im$, i.e. $\left(m^*\right)^2 = -m^2 < 0$.

4.3. On The Possible Void Decay in Free – Electron Laser SASE – FEL Experiment

We must make some profound alterations to the theoretical idea of the vacuum. . . . Thus, with the new theory of electrodynamics we are rather forced to have an aether.

P.A.M. Dirac, Nature, 1951, vol. 168, pp. 906-907

Recently the ultra-high energy lasers proposals are developed [4.5]. The SASE-FEL project for the first time enables the investigation of the "vacuum decay" processes and emission of ultra-relativistic fermions. In this chapter we argue that the results of the SASE-FEL future experiments open the new field of the investigation of the structure of the spacetime.

Bell's theorem is rooted in two assumptions: the objective reality – the reality of the external world, independent of our observations; the second is locality, or no faster than light signaling. Aspect's experiment appears to indicate that one of these two has to go.

In this paragraph we are going back to relativity as it was before Einstein when people like Lorentz and Poincaré thought that there was an aether – a preferred frame of reference – but that our measuring instruments were distorted by motion in such a way that we could not detect motion through the aether. Now in that way you can imagine that there is a preferred frame of reference and in this preferred frame of reference particles do go faster than light. But then in other, our, frame of reference particles have the speed lower than the light speed.

In this chapter we propose the following scenario. Behind the scene – our world of observation something is going which is not allowed to appear on the scenes.

To start with we observe that for electrons, at the vicinity of $E_k \sim mc^2$ the speed of electrons is changed abruptly. With $E_k \sim mc^2$, through Heisenberg inequality the characteristic time can be defined

$$\tau = \frac{\hbar}{mc^2}. \tag{4.32}$$

At that characteristic time the Newton theory (NT) and SR theory starts to give different description of the speeds, viz.,

$$\upsilon_{NT} = \sqrt{\frac{2E_k}{m_k}} \tag{4.33}$$

and

$$\upsilon_{SR} = c\sqrt{1 - \frac{1}{\left(\frac{E_k}{mc^2} + 1\right)^2}}. \tag{4.34}$$

Let us introduce the acceleration a_m which describes the change of speeds in time τ

$$a_m = \frac{\upsilon_{NT} - \upsilon_{SR}}{\tau}. \tag{4.35}$$

and force F which opposes the acceleration of the particle with mass m

$$F = ma_m = \frac{mc}{\tau}\left(\left(\frac{2E_k}{mc^2}\right)^{\frac{1}{2}} - \left(1 - \frac{1}{\left(\frac{E_k}{mc^2}+1\right)^2}\right)^{\frac{1}{2}}\right)$$
$$= \frac{m^2c^3}{\hbar}\left(\left(\frac{2E_k}{mc^2}\right)^{\frac{1}{2}} - \left(1 - \frac{1}{\left(\frac{E_k}{mc^2}+1\right)^2}\right)^{\frac{1}{2}}\right). \tag{4.36}$$

In formula (4.36) we introduce the field E_s

$$E_s = \frac{m^2c^3}{\hbar e} \tag{4.37}$$

and we obtain:

$$F = E_s e = \left(\frac{2E_k}{mc^2}\right)^{\frac{1}{2}} - \left(1 - \frac{1}{\left(\frac{E_k}{mc^2}+1\right)^2}\right)^{\frac{1}{2}}. \tag{4.38}$$

It is interesting to observe that the field E_s is the same as the Schwinger field strengths [4.6]. J. Schwinger demonstrated that in the background of a static electric field, the QED vacuum is broken and decayed with spontaneous emission of e^+e^- pairs.

In the following we define for electrons the energy L

$$L = eF_s r_e = \alpha\frac{\left(m_e c^2\right)^4}{e^6}. \tag{4.39}$$

In the formula (4.39) r_e is the classical electron radius

$$r_e = \frac{e^2}{m_e c^2} \tag{4.40}$$

and α is the fine structure constant. Having the energy L and volume r_e^3 where the energy is concentrated we define the bulk modulus for the medium which opposes the motion of electron

$$B = \frac{L}{r_e^3} = \alpha \frac{(m_e c^2)^4}{e^6} \tag{4.41}$$

and hypothetic sound velocity in the medium which oppose the acceleration of the electrons

$$\upsilon_{sound} = \left(\frac{B}{\rho}\right)^{\frac{1}{2}} = 10^{18} c, \tag{4.42}$$

where $\rho = \frac{m}{r_e^3}$.

4.4. Fourier Diffusion and Special Relativity in Nanotechnology

The nanotechnology industry has evolved at very high rate, particularly in the past ten years. Transistor channel length has decreased from 2.0 μm in 1980 to 0.5 μm in 1992 to current (2006) systems with channel length ~ 50nm. Recently the International Technology Road Map for Semiconductors (ITRS) was postulated. The IRST is devoted to the study of the proposed of the global interconnects and problems with the time delays of the transistors and logical systems. One of the *remedium* for system time delays is the technology On – Chip Transmission Lines. The On – Chip Transmission Technology allows for the electron transmission with near – of – light On – Chip electrical interconnection

This chapter is addressed to the thermal transport in On – Chip – Transmission Lines.

In the first paragraph we develop the theoretical framework for heat transport in On – Chip – Transmission Line (OCTL). We formulate the transmission line heat transport equation and solved it for the Cauchy boundary conditions.

In the second paragraph we study the special relativity influence on the information transmission in OCTL. We show that the standard Fourier approximation leads to infinite speed of the data transmission. On the other hand the hyperbolic diffusion equation formulated in paragraph 1 gives the finite speed of the transmission of the data.

Dynamics of nonequilibrium electrons and phonons in metals, semiconductors have been the focus of much attention because of their fundamental interest in solid state physics and nanotechnology.

In metals, relaxation dynamics of optically excited nonequilibrium electrons has been extensively studied by pump – probe techniques using femtosecond lasers [4.7 – 4.10].

Recently [4.11] it was shown that the optically excited metals relax to equilibrium with two models: rapid electron relaxation and slow thermal relaxation through the creation of the optical phonons. The same processes will occur in OCTL

In this chapter we develop the hyperbolic thermal diffusion equation with two models: electrons and phonons relaxation. These two modes are characterized by two relaxation times τ_1 for electrons and τ_2 for phonons. This new equation is the generalization of our one mode hyperbolic equation, with only electrons degrees of freedom, τ_1 [4.12]. The hyperbolic two mode equation is the analogous equation to Klein – Gordon equation and allows the heat propagation with finite speed.

As was shown in paper [4.12] for high frequency laser pulses the diffusion velocity exceeds that of light. This is not possible and merely demonstrates that Fourier equation is not really correct. Oliver Heaviside was well aware of this writing [4.13]:

> All diffusion formulae (as in heat conduction) show instantaneous action to the infinite distances of a source, though only to an infinitesimal extent. To make the theory of heat diffusion be rational as well as practical some modification of the equation to remove the instantaneity, however little difference it may make quantatively, in general.

August 1876 saw the appearance in Philosophical Magazine [4.13] the paper which extended the mathematical understanding the diffusion (Fourier) equation. O. Heaviside for the first time wrote the hyperbolic diffusion equation for the voltage $V(x, t)$. Assuming a uniform resistance, capacitance and inductance per unit length, k, c and s respectively he arrived at:

$$\frac{\partial^2 V(x,t)}{\partial x^2} = sc\frac{\partial^2 V(x,t)}{\partial t^2} + kc\frac{\partial V(x,t)}{\partial t}. \quad (4.43)$$

The discussion of the broad sense of the Heaviside equation (4.43) can be find out, for example in our monograph [4.2], viz,

$$\tau^2\frac{\partial^2 T}{\partial t^2} + \tau\frac{\partial T}{\partial t} + \frac{2V\tau}{\hbar}T = \tau\frac{\hbar}{m}\nabla^2 T. \quad (4.44)$$

Equation (4.44) is the heat transport equation for the transmission line. In Eq. (4.44) $T(\vec{r},t)$ denotes the temperature field, V is the external potential, m is the mass of heat carrier and τ is the relaxation time

$$\tau = \frac{\hbar}{mv^2}. \quad (4.45)$$

As can be seen from formulae (4.44) and (4.45) in hyperbolic diffusion equation the same relaxation time τ is assumed for both type of motion: wave and diffusion.

This can not be so obvious. For example let us consider the simpler harmonic oscillator equation:

$$m\frac{d^2x}{dt^2}+kx+c\frac{dx}{dt}=0. \tag{4.46}$$

Equation (4.46) can be written as

$$\tau^2\frac{d^2x}{dt^2}+\tau\frac{dx}{dt}+x=0, \tag{4.47}$$

where

$$\tau^2=\frac{m}{k}, \qquad \tau=\frac{c}{k} \tag{4.48}$$

i.e.

$$c^2=km. \tag{4.49}$$

As it was well known equation (4.47) with formula (4.48) describes only the weakly damped (periodic) motion of the harmonic oscillator (HO). It must be stressed that for HO exists also critically damped and overdamped modes which are not describes by the equation (4.47). The general master equation for HO must be written as

$$\tau_1^2\frac{d^2x}{dt^2}+\tau_2\frac{dx}{dt}+x=0. \tag{4.50}$$

Following the discussion of the formulae (4.47) to (4.50) we argue that the general hyperbolic diffusion equation can be written as:

$$\tau_1^2\frac{\partial^2T}{\partial t^2}+\tau_2\frac{\partial T}{\partial t}+\frac{2V\tau_2}{\hbar}T=\tau_2\frac{\hbar}{m}\nabla^2T \tag{4.51}$$

and $\tau_1\neq\tau_2$.

Equation (4.51) describes the temperature field generated by ultra-short laser pulses. In Eq. (4.51) two modes: wave and diffusion are described by different relaxation times.

$$T(x,t)=e^{-\frac{t\tau_2}{2\tau_1^2}}u(x,t). \tag{4.52}$$

After substitution Eq. (4.52) into Eq. (4.51) one obtains

$$\frac{1}{\upsilon^2}\frac{\partial^2 u}{\partial t^2}-\frac{\partial^2 u}{\partial x^2}+q^2u(x,t)=0, \tag{4.53}$$

where

$$\upsilon^2=\frac{\hbar\tau_2}{m\tau_1^{\ 2}},\quad q^2=\left(\frac{2Vm}{\hbar^2}-\frac{1}{4}\frac{m}{\hbar}\frac{\tau_2}{\tau_1^{\ 2}}\right). \tag{4.54}$$

Equation (4.53) is the thermal two-mode Klein – Gordon equation and is the generalization of Klein – Gordon one mode equation developed in our monograph.

For Cauchy initial condition

$$u(x,0)=f(x),\qquad u_t(x,0)=g(x) \tag{4.55}$$

the solution of Eq. (4.53) has the form

$$\begin{aligned}u(x,t)&=\frac{f(x-\upsilon t)+f(x+\upsilon t)}{2}\\&+\frac{1}{2\upsilon}\int_{x-\upsilon t}^{x+\upsilon t}g(\varsigma)I_0\left[\sqrt{-q^2\left(\upsilon^2t^2-(x-\varsigma)^2\right)}\right]d\varsigma\\&+\frac{\upsilon\sqrt{-q^2}t}{2}\int_{x-\upsilon t}^{x+\upsilon t}f(\varsigma)\frac{I_1\left[\sqrt{-q^2\left(\upsilon^2t^2-(x-\varsigma)^2\right)}\right]}{\sqrt{\upsilon^2t^2-(x-\varsigma)^2}}d\varsigma\end{aligned} \tag{4.56}$$

for $q^2<0$

and

$$\begin{aligned}u(x,t)&=\frac{f(x-\upsilon t)+f(x+\upsilon t)}{2}\\&+\frac{1}{2\upsilon}\int_{x-\upsilon t}^{x+\upsilon t}g(\varsigma)J_0\left[\sqrt{q^2\left(\upsilon^2t^2-(x-\varsigma)^2\right)}\right]d\varsigma\\&+\frac{\upsilon\sqrt{q^2}t}{2}\int_{x-\upsilon t}^{x+\upsilon t}f(\varsigma)\frac{J_1\left[\sqrt{q^2\left(\upsilon^2t^2-(x-\varsigma)^2\right)}\right]}{\sqrt{\upsilon^2t^2-(x-\varsigma)^2}}d\varsigma\end{aligned} \tag{4.57}$$

for $q^2>0$.

In this paragraph we develop the description of the heat transport in Minkowski spacetime of the OCTS. In the context special relativity theory we investigate the Fourier equation and hyperbolic diffusion equation. We calculate the speeds of the heat diffusion in Fourier approximation and show that for high energy laser beam the heat diffusion exceeds the light velocity. We show that this result breaks the causality of the thermal phenomena in Minkowski spacetime. The same phenomena we describe in the framework of hyperbolic heat diffusion equation and show that in that case speed of diffusion is always smaller than light velocity. We may use the concept that the speed of light *in vacuo* provides an upper limit on the speed with which a signal can travel between two events to establish whether or not any two events could be connected. In the interest of simplicity we shall work with one space dimension $x_1 = x$ and the time dimension $x_o = ct$ of the Minkowski spacetime. Now let us consider events (1) and (2): their Minkowski interval Δs satisfies the relationship:

$$\Delta s^2 = c^2 \Delta t^2 - \Delta x^2 . \tag{4.58}$$

Without loss of generality we take Event 1 to be at $x = 0$, $t = 0$. Then Event 2 can be only related to Event 1 if it is possible for a signal traveling at the speed of light, to connect them. Let then Event 2 is at (Δx, $c\Delta t$), its relationship to Event 1 depending on whether $\Delta s > 0$, $= 0$, or < 0.

We may summarize the three possibilities as follows:

Case A *timelike interval,* $|\Delta x_A| < c\Delta t$, or $\Delta s^2 > 0$. Event 2 can be related to Event 1, events 1 and 2 can be in causal relation.

Case B *lightlike interval,* $|\Delta x_B| = c\Delta t$, or $\Delta s^2 = 0$. Event 2 can only be related to Event 1 by a light signal.

Case C *spacelike interval* $|\Delta x_A| > c\Delta t$, or $\Delta s^2 < 0$. Event 2 cannot be related to Event 1, for in that case $\upsilon > c$.

Now let us consider the case C in more details. At first sight it seems that in case C we can find out the reference frame in which two Events $c^>$ and $c^<$ always fulfils the relations $t_{c^>} - t_{c^<} > 0$. But it is not true. Let us choose the inertial frame U' in which $t_{c^>} - t_{c^<} > 0$. In reference frame U which is moving with speed V relative to U', where

$$V = c\frac{c(t'_{c^>} - t'_{c^<})}{x'_{c^<} - x'_{c^>}} \tag{4.59}$$

Speed $V < c$ for

$$\left|\frac{c(t'_{c^>} - t'_{c^<})}{x'_{c^<} - x'_{c^>}}\right| < 1 \tag{4.60}$$

Let us calculate $t_{c^>} - t_{c^<}$ in the reference frame U

$$t_{c^>} - t_{c^<} = \frac{1}{\sqrt{1-\frac{V^2}{c^2}}}\left[\frac{V}{c^2}(x'_{c^>} - x'_{c^<}) + (t'_{c^>} - t'_{c^<})\right] =$$

$$\frac{1}{\sqrt{1-\frac{V^2}{c^2}}}\left[\frac{t'_{c^>} - t'_{c^<}}{x'_{c^>} - x'_{c^<}}(x'_{c^>} - x'_{c^<}) + (t'_{c^>} - t'_{c^<})\right] = 0 \tag{4.61}$$

For the greater V we will have $t_{c^>} - t_{c^<} < 0$. It means that for the spacelike intervals the sign of $t_{c^>} - t_{c^<}$ depends on the speed V, i.e. causality relation for spacelike events is not valid.

In paper [4.14] the speed of the diffusion signals was calculated

$$\upsilon = \sqrt{2D\omega} \tag{4.62}$$

where

$$D = \frac{\hbar}{m} \tag{4.63}$$

and ω is the angular frequency of the laser pulses. Considering formula (4.62) and (4.63) one obtains

$$\upsilon = c\sqrt{2\frac{\hbar\omega}{mc^2}} \tag{4.64}$$

and $\upsilon \geq c$ for $\hbar\omega \geq mc^2$.

From formula (4.64) we conclude that for $\hbar\omega > mc^2$ the Fourier diffusion equation is in contradiction with special relativity theory and breaks the causality in transport phenomena.

In monograph [4.2] the hyperbolic model of the heat transport phenomena was formulated. It was shown that the description of the ultrashort thermal energy transport needs the hyperbolic diffusion equation (one dimension transport)

$$\tau\frac{\partial^2 T}{\partial t^2} + \frac{\partial T}{\partial t} = D\frac{\partial^2 T}{\partial x^2}. \tag{4.65}$$

In the equation (4.65) $\tau = \dfrac{\hbar}{m\alpha^2 c^2}$ is the relaxation time, m = mass of the heat carrier, α is the coupling constant and c is the light speed in vacuum, $T(x,t)$ is the temperature field and $D = \hbar/m$.

In paper [4.14] the speed of the thermal propagation υ was calculated

$$\upsilon = \frac{2\hbar}{m}\sqrt{-\frac{m}{2\hbar}\tau\omega^2 + \frac{m\omega}{2\hbar}(1+\tau^2\omega^2)^{\frac{1}{2}}} . \tag{4.66}$$

Considering that $\tau = \hbar/m\alpha^2 c^2$ formula (4.66) can be written as

$$\upsilon = \frac{2\hbar}{m}\sqrt{-\frac{m}{2\hbar}\frac{\hbar\omega^2}{mc^2\alpha^2} + \frac{m\omega}{2\hbar}(1+\frac{\hbar^2\omega^2}{m^2c^4\alpha^4})^{\frac{1}{2}}} . \tag{4.67}$$

For

$$\frac{\hbar\omega}{mc^2\alpha^2} < 1, \qquad \frac{\hbar\omega}{mc^2} < 1 \tag{4.68}$$

one obtains from formula (4.67)

$$\upsilon = \sqrt{\frac{2\hbar}{m}\omega} . \tag{4.69}$$

Formally formula (4.69) is the same as formula (4.64) but considering inequality (4.68) we obtain

$$\upsilon = \sqrt{\frac{2\hbar\omega}{m}} = \sqrt{2}\alpha c < c \tag{4.70}$$

and causality is not broken.

For

$$\frac{\hbar\omega}{mc^2} > 1; \quad \frac{\hbar\omega}{\alpha^2 mc^2} > 1 \tag{4.71}$$

we obtain from formula (4.67)

$$\upsilon = \alpha c,\ \upsilon < c. \tag{4.72}$$

Considering formulae (4.70) and (4.72) we conclude that the hyperbolic diffusion equation (4.65) describes the thermal phenomena in accordance with special relativity theory and causality is not broken independently of laser beam energy.

4.5. Proca Equation for Attosecond Electron Pulses

In the seminal paper [4.15] prof. Ahmed Zewail and P. Baum put forward the very interesting idea of the creation of the attosecond electron pulses. In this chapter we develop the theoretical model for the heat transport of the attosecond electron pulses. As can be shown the electron heated by attosecond electron pulses can evolve as the heat wave, due to hyperbolicity of the master equation.

Parabolic theories of dissipative phenomena have long and a venerable history and proved very useful, especially in the steady-state regime [4.2]. They exhibit, however, some undesirable features, such as a causality (see e.g., [4.2]), which prompted the formulation of hyperbolic theories of dissipation to get rid of them. This was achieved at a price by extending the set of field variables by including the dissipative fluxes (heat current, non-equilibrium stresses and so on) on the same footing as the classical (energy densities, equilibrium pressures, etc), thereby giving rise to a set of 'more physically' satisfactory (as they conform much better with experiments) but involved theories from the mathematical point of view. These theories have the additional advantage of being backed by statistical fluctuation theory, the kinetic theory of gases (Grad's 13-moment approximation), information theory and correlated random walks.

A key quantity in these theories is the relaxation time W of the corresponding dissipative process. This positive-definite quantity has a distinct physical meaning, namely the time taken by the system to return spontaneously to the steady state (whether of thermodynamic equilibrium or not) after it has been suddenly removed from it. It is, however, connected to the mean collision time t_c of the particles responsible for the dissipative process It is therefore appropriate to interpret the relaxation time as the time taken by the corresponding dissipative flow to relax to its steady value.

In the book [4.2] the new hyperbolic Proca type equation for heat transport was formulated and solved.

The excitation of matter on the atomic level leads to transfer of energy. The response of the matter is governed by the relaxation time.

In this paragraph we develop the general, universal definition of the relaxation time, which depends on coupling constants for electromagnetic interaction.

It occurs that the general formula for the relaxation time can be written as

$$\tau_i = \frac{\hbar}{m_i(\alpha_i c)^2}, \tag{4.73}$$

where m_i is the heat carrier mass, $\alpha_i = \left(i = e, \frac{1}{137}\right)$ is coupling constant for electromagnetic interaction, c is the vacuum light speed. As the c is the maximum speed all relaxation time fulfils the inequality

$$\tau > \tau_i .$$

Consequently τ_i is the minimal universal relaxation time.

After the standards of time and space were defined, the laws of classical physics relating such parameters as distance, time, velocity, temperature are assumed to be independent of the accuracy with which, these parameters can be measured. It should be noted that this assumption does not enter explicitly into the formulation of classical physics. It implies that together with the assumption of existence of an object and really independently of any measurements (in classical physics) it was tacitly assumed that *there was a possibility of an unlimited increase in the accuracy of measurements.* Bearing in mind the "atomicity" of time i.e. considering the smallest time period, the Planck time, the above statement is obviously not true. Atto-second electron pulses are at the limit of time resolution.

In this paragraph, we develop and solve the quantum relativistic heat transport equation for attosecond electron transport phenomena, where external forces exist [4.2]. We develop the new hyperbolic heat transport equation which generalises the Fourier heat transport equation for the rapid thermal processes have been written in the form:

$$\frac{1}{\left(\frac{1}{3}\upsilon_F^2\right)}\frac{\partial^2 T}{\partial t^2} + \frac{1}{\tau\left(\frac{1}{3}\upsilon_F^2\right)}\frac{\partial T}{\partial t} = \nabla^2 T , \tag{4.74}$$

where T denotes the temperature, τ the relaxation time for the thermal disturbance of the fermionic system, and υ_F is the Fermi velocity.

In what follows we present the formulation of the HHT, considering the details of the two fermionic systems: electron gas in metals.

For the electron gas in metals, the Fermi energy has the form

$$E_F^e = (3\pi)^2 \frac{n^{2/3}\hbar^2}{2m_e}, \tag{4.75}$$

where n denotes the density and m_e electron mass. Considering that

$$n^{-1/3} \sim a_B \sim \frac{\hbar^2}{me^2}, \tag{4.76}$$

and a_B = Bohr radius, one obtains

$$E_F^e \sim \frac{n^{2/3}\hbar^2}{2m_e} \sim \frac{\hbar^2}{ma^2} \sim \alpha^2 m_e c^2, \tag{4.77}$$

where c = light velocity and $\alpha = 1/137$ is the fine-structure constant for electromagnetic interaction. For the Fermi momentum p_F we have

$$p_F^e \sim \frac{\hbar}{a_B} \sim \alpha m_e c, \tag{4.78}$$

and, for Fermi velocity υ_F,

$$\upsilon_F^e \sim \frac{p_F}{m_e} \sim \alpha c. \tag{4.79}$$

Considering formula (4.79), equation HHT can be written as

$$\frac{1}{c^2}\frac{\partial^2 T}{\partial t^2} + \frac{1}{c^2 \tau}\frac{\partial T}{\partial t} = \frac{\alpha^2}{3}\nabla^2 T. \tag{4.80}$$

As seen from (4.80), the HHT equation is a relativistic equation, since it takes into account the finite velocity of light.

In the following, the procedure for the quantisation of temperature $T(r,t)$ in a hot fermion gas will be developed. First of all, we introduce the reduced de Broglie wavelength

$$\lambda_B^e = \frac{\hbar}{m_e \upsilon_h^e}, \qquad \upsilon_h^e = \frac{1}{\sqrt{3}}\alpha c \tag{4.81}$$

and the mean free path λ^e

$$\lambda^e = \upsilon_h^e \tau^e. \tag{4.82}$$

In view of the equations (4.81) and (4.82), we obtain the HHC for electron and nucleon gases

$$\frac{\lambda_B^e}{\upsilon_h^e}\frac{\partial^2 T}{\partial t^2} + \frac{\lambda_B^e}{\lambda^e}\frac{\partial T}{\partial t} = \frac{\hbar}{m_e}\nabla^2 T^e. \tag{4.83}$$

Equation (4.83) is the hyperbolic partial differential equation which is the master equation for heat propagation in Fermi electron and nucleon gases. In the following, we study the quantum limit of heat transport in the fermionic systems. We define the quantum heat transport limit as follows:

$$\lambda^e = \lambda_B^e. \tag{4.84}$$

In that case, Eq. (4.83) has the form

$$\tau^e \frac{\partial^2 T^e}{\partial t^2} + \frac{\partial T^e}{\partial t} = \frac{\hbar}{m_e}\nabla^2 T^e, \tag{4.85}$$

where

$$\tau^e = \frac{\hbar}{m_e\left(\upsilon_h^e\right)^2}. \tag{4.86}$$

Equations (4.85) and (4.86) define the master equation for quantum heat transport (QHT). With the relaxation relaxation time τ^e, one can define the "pulsation" ω_h^e

$$\omega_h^e = (\tau^e)^{-1}, \tag{4.87}$$

or

$$\omega_h^e = \frac{m_e\left(\upsilon_h^e\right)^2}{\hbar},$$

i.e.,

$$\omega_h^e\hbar = m_e\left(\upsilon_h^e\right)^2 = \frac{m_e\alpha^2}{3}c^2. \tag{4.88}$$

The equations (4.88) define the Planck-Einstein relation for heat quanta E_h^e:

$$E_h^e = \omega_h^e\hbar = m_e\left(\upsilon_h^e\right)^2. \tag{4.89}$$

The heat quantum with energy $E_h = \hbar\omega$ can be named the *heaton*, in complete analogy to the *phonon*, *magnon*, *roton*, etc. For $\tau^e, \tau^N \to 0$, Eq. (4.85) is the Fourier equation with quantum diffusion coefficient D^e

$$\frac{\partial T^e}{\partial t} = D^e \nabla^2 T^e, \qquad D^e = \frac{\hbar}{m_e}. \tag{4.90}$$

For finite τ^e, for $\Delta t < \tau^e$, Eq. (4.85) can be written as

$$\frac{1}{(\upsilon_h^e)^2} \frac{\partial^2 T^e}{\partial t^2} = \nabla^2 T^e. \tag{4.91}$$

Equations (4.90) and (4.91) are the wave equations for quantum heat transport (QHT)

It is quite interesting that the Proca type equation can be obtained for thermal phenomena. In the following, starting with the hyperbolic heat diffusion equation the Proca equation for thermal processes will be developed [4.2].

When the external force is present $F(x,t)$ the forced damped heat transport is obtained [4.2] (in a one dimensional case):

$$\frac{1}{\upsilon^2} \frac{\partial^2 T}{\partial t^2} + \frac{m_0 \gamma}{\hbar} \frac{\partial T}{\partial t} + \frac{2Vm_0\gamma}{\hbar^2} T - \frac{\partial^2 T}{\partial x^2} = F(x,t). \tag{4.92}$$

The hyperbolic relativistic quantum heat transport equation, (4.92), describes the forced motion of heat carriers, which undergo scattering ($\frac{m_0 \gamma}{\hbar} \frac{\partial T}{\partial t}$ term) and are influenced by the potential term ($\frac{2Vm_0\gamma}{\hbar^2} T$).

Equation (4.92) is the Proca thermal equation and can be written as [4.2]:

$$\left(\overline{\square}^2 + \frac{2Vm_0\gamma}{\hbar^2} \right) T + \frac{m_0\gamma}{\hbar} \frac{\partial T}{\partial t} = F(x,t),$$
$$\overline{\square}^2 = \frac{1}{\upsilon^2} \frac{\partial^2}{\partial t^2} - \frac{\partial^2}{\partial x^2}. \tag{4.93}$$

We seek the solution of equation (4.93) in the form

$$T(x,t) = e^{-t/2\tau} u(x,t), \tag{4.94}$$

where $\tau = \frac{\hbar}{m\upsilon^2}$ is the relaxation time. After substituting equation (4.94) in equation (4.93) we obtain a new equation

$$\left(\overline{\square}^2 + q^2 \right) u(x,t) = e^{t/2\tau} F(x,t) \tag{4.95}$$

and

$$q^2 = \frac{2Vm}{\hbar^2} - \left(\frac{mv}{2\hbar}\right)^2 \tag{4.96}$$

$$m = m_0\gamma . \tag{4.97}$$

In free space i.e. when $F(x,t) \to 0$ equation (4.93) reduces to

$$\left(\bar{\Box}^2 + q^2\right)u(x,t) = 0 , \tag{4.98}$$

which is essentially the free Proca type equation.

The Proca equation describes the interaction of the atto-second electron pulse with the matter. As was shown in book [4.2] the quantization of the temperature field leads to the *heatons* – quanta of thermal energy with a mass $m_h = \frac{\hbar}{\tau v_h^2}$, where τ is the relaxation time and v_h is the finite velocity for heat propagation. For $v_h \to \infty$, i.e. for $c \to \infty$, $m_0 \to 0$, it can be concluded that in non-relativistic approximation (c = infinite) the *Proca* equation is the diffusion equation for mass-less photons and heatons.

REFERENCES

[4.1]. Ott, H. *Z. Phys*. 1963, *175*, 70.
[4.2]. Kozlowski, M.; Marciak – Kozlowska, J. *Thermal Processes Using Attosecond Laser Pulses*, Optical Science 121; Springer: New York, NY, 2006.
[4.3]. Chang, T. *Nucl. Sci. Tech.* 2002, *13*, 129.
[4.4]. Ai, X.; Li, B.Q. *Journal of Elec. Mat.* 2005, *34*, 583.
[4.5]. TESLA, SASE-FEL Technical Design Project.
[4.6]. Schwinger, J. *Phys. Rev*. 1951, *82,* 664.
[4.7]. Sun C. K.; et al. *Phys. Rev*. 1993, *B48,* R 12365.
[4.8]. Fan W. S.; et al. *Phys. Rev*. 1992, *B46,* 13592.
[4.9]. Del Fatti, M.; et al. *Phys. Rev. Lett*. 1998, *81,* 922.
[4.10]Hohlfeld, J.; et al. *Chem. Phys*. 2000, *251,* 237.
[4.11]Hase, M.; et al. (2005). Ultrafast dynamics of coherent and nonequilibric electrons in transition metals. arXiv:cond-mat/0504540.
[4.12]Marciak – Kozlowska, J.; Kozlowski, M. *Laser In Engineering*, 2006, *16*, 10.
[4.13.]Heaviside, O. *Electromagnetic Theory*; Chelsea Publishing Company: New York, NY, 1971; Vol. 2.
[4.14]Marciak – Kozlowska, J.; Kozlowski, M. *Lasers in Engineering,* 2006, *16*, 140.
[4.15]Baum, P.; Zewail, A. *PNAS* 2007, *104*, 18409-18410.

Chapter 5

Extreme Light, Lasetron and Tesla SASE – FEL

5.1. Space – Time Engineering with LASETRON Pulses

Ever shorter electromagnetic (EM) pulses have always been of keen interest, largely as a means of investigating and controlling ever faster processes. Recent proposals [5.1] explored various avenues to attaining the shortest, subfemtosecond (10^{-16} – 10^{-17} s), EM pulses of atomic time-scale duration. In the most recent breakthrough work [5.2], the train of ~ 0.25 fs pulses has been observed experimentally. The further scale of fundamental interest is that of strong nuclear interactions. Estimating the time scale of a process as $\hbar$/(energy scale), and keeping in mind that atomic energy scale, defined as the atomic ionization limit, is ~ 10 – 20 eV, while the nuclear energies are beyond 1 MeV, one finds that the nuclear time scale is shorter by about 5 orders of magnitude, i.e., in the 10^{-21} – 10^{-22} s domain [zeptosecond (zs) to sub-zs]. The feasibility of such pulses has not been considered yet.

In this paragraph the theoretical discussion of the zs and sub-zs pulses is presented. It will be shown that ultra-short laser pulses can be generated using petawatt lasers, while already available terawatt lasers may generate subattosecond pulses of ~ 10^{-19} s. The pulses will be radiated by ultrarelativistic electrons driven by circularly polarized high-intensity laser fields. They are basically reminiscent to synchrotron radiation; no synchrotron, however, can even come close to running electrons with the energy of 50 MeV at the (laser) frequency $\omega_L \sim 10^{15}$ – 10^{16} s^{-1} in the 0.1 μm radius orbit, as a petawatt laser can. The major distinct feature here is the forced synchronization of all radiating electrons by the driving laser field. Radiation of such a synchronized bunch would be viewed by an observer at any point in the rotation plane as huge pulses/bursts of EM field as short as

$$\tau_{pl} \sim \frac{1}{2\omega_L \gamma^3}, \tag{5.1}$$

where γ is the electron's relativistic factor. With $\lambda_L = 2\pi c/\omega_L \sim 1$ μm and $\gamma \sim 64$ (attainable with a petawatt laser), we have $\tau_{pl} \approx 10^{-21}$ s. We call such a system LASETRON. It can be achieved by placing a solid particle or a piece of wire of subwavelength cross section in the focal plane of a superpowerful laser. In addition to zeptosecond pulses with substantial energy, the magnetic field at the center of rotation may reach ~ 10^6 T – comparable to fields in the vicinity of white dwarves. Our results also show that the coherent radiation friction drastically limits the rotation energy of electrons in ultraintense laser fields.

Relativistically intense laser field interactions with free electrons have been extensively explored (see, e.g., Ref. [5.3] and references therein). Closer to our subject, [5.4] suggests using electrons in high-intensity, circularly polarized laser light to generate high-order harmonics. However, coherent generation of high-intensity ultra-short pulses by free electrons in laser fields, especially when the radiated power is so high that the radiation damping would dramatically affect the electron motion, has not, to our knowledge, been addressed thus far.

It is known (see, e.g., [5.5]) that in a circularly polarized EM pulse an electron moves in a helix. We neglect the longitudinal motion by considering only a quasi-steady-state situation when the electron moves in a circle for the time period much longer than one laser cycle. This simple, while advantageous, configuration might be created, for instance, by two circularly polarized counterpropagating laser pulses [5.6]. Recent results [5.7] on a similar configuration with a thin film target predict the formation of a very thin layer of free electrons with the density $n_e \sim 10^{23}$ cm^{-3}.

A circularly polarized laser field with amplitude E_L drives an electron with the speed υ in a circle with the radius $\rho = \upsilon\lambda_L / 2\pi c$ in phase with the field, so that its momentum $p = \upsilon\gamma m_e$ and relativistic factor $\gamma = \left(1-\beta^2\right)^{-1/2}$, where $\beta = \upsilon/c$, are, respectively, as [5.8]

$$p = \varepsilon_L m_e c, \qquad \gamma = \sqrt{1+\varepsilon_L^2}, \qquad \varepsilon_L = E_L / E_{rel}, \tag{5.2}$$

where $E_{rel}(\omega_L) = m_e \omega_L c / e \approx 10^4 \lambda_L^{-1}$ esu is a relativistic scale of the field strength. (We will see below, however, that Eq. (5.2) should be amended for high radiation losses.) A rotating electron will radiate the power [5.6, 5.8] P_e

$$P_e = m_e c^2 \omega_L \Gamma_e \gamma^2 \left(\gamma^2 - 1\right), \qquad \Gamma_e = \frac{4\pi r_e}{3\lambda_L}, \tag{5.3}$$

where $r_e = e^2/m_e c^2 \approx 2{,}82 \times 10^{-13}$ cm is the classical electron radius. The radiation is concentrated in the angle $\theta \sim 1/\gamma$ around the direction of the instantaneous electron velocity, and an observer in the rotation plane may see only ultra-short bursts of radiation of the duration τ_{pl}, Eq. (5.1), separated by the laser period, $2\pi/\omega_L$. The Fourier spectrum of the bursts spreads up to the (classical) cutoff

$$\omega_{\max} \sim 3\gamma^3 \omega_L. \tag{5.4}$$

Quantum cutoff frequency, that of the photon with the energy equal to the electron rotation energy, is $\omega_{qm} \sim m_e \times c^2(\gamma - 1)/\hbar$. They coincide at

$\gamma \sim \left(m_e c^2 / 3\hbar\omega_L\right)^{1/2} = O(10^3)$; therefore, even for a petawatt laser with γ ~ 100 (see below), the radiation near ω_{qm} is negligible.

We estimate the parameters of the lasetron radiation for coherent radiation sources focused tightly to a few-wavelength spot size $w_L = j\lambda_L$, $j = O(1)$. One has then $\gamma^2 \approx 1 + 4.6\times10^{-11} P_L / j^2$, where P_L is the driving laser power. For simplicity, we assume $j^2 \approx 4.6$ (the beam waist area is then $\approx 7\lambda_L^2$), so that

$$\gamma^2 = 1 + 10P_L(\text{TW}). \tag{5.5}$$

For quantitative estimates, we consider the following model sources: (i) $PL - P_L = 10^{15}$ W (petawatt) laser at $\lambda_L = 1$ μm, a close approximation to the LLNL petawatt laser and a similar system under construction in Japan; (ii) MTW (multiterawatt) – a 100 TW CO_2, $\lambda_L = 10$ μm system under construction in Japan; the CO_2 laser at 40 TW is in operation; (iii) LTW—quite widespread lasers of few-TW power; as an example, we will use a 5 TW system at $\lambda_L = 0.8$ μm; and (iv) RK—a relativists klystron under development at NRL, potentially a 1 TW system with $\lambda_L \approx 3$ cm. Table 5.1 illustrates lasetron radiation parameters for a single electron for all these sources. Thus, a single electron in the focus of the petawatt laser would radiate a macroscopic power of 180 W in nuclear time-scale bursts, $\tau_{pl} = 0.26$ zs. The classical cutoff, $\hbar\omega_{cl} = 3$ MeV, lies above the energy threshold of some *photonuclear reactions,* e.g., neutron photoproduction on Be (1.7 MeV). This indicates the potential of lasetron for time-resolved photonuclear physics—provided that a burst carries sufficient energy.

Table 5.1. Single-Electron Output of Model Sources

Source	Pe	τpl	$\hbar\omega_{\max}$	Burst
PL	180 W	0.26 zs	3.9 MeV	0.3 eV
MTW	0.02 W	0.81 as	13 keV	0.1 eV
LTW	0.6 μW	0.6 as	1.7 keV	1.9 μeV
RK	1.5 pW	220 fs	5 meW	0.3 μeV

Unfortunately, even for PL, the energy radiated by one electron in one burst is still very low. To increase it substantially, many electrons have to radiate coherently. A straightforward solution is to place a sub-λ_L, size solid particle in the focal plane of a high-power laser, which would fully ionize the particle within a fraction of a laser cycle. Free electrons will then experience an "orbital sander" rotation, moving in phase with the field in identical but shifted circular orbits, their relative positions fixed. The resulting radiation will be almost fully coherent, with the ranted power scaling as the particle number squared,

$$P_{rad} \approx N_e^2 P_e. \tag{5.6}$$

Now, however, P_e cannot be taken from Eqs. (5.2) and (5.3), because we have to take into account a new factor—*coherent* radiation friction, or back-reaction of radiation, which is a major player in the phenomenon we consider. Indeed, applying Eq. (5.6), with γ for P_e taken from Eq. (5.2), for N_e as small as 10^7, one obtains P_{rad} = 18 PW—much higher than the full driving power. The reason for this contradiction is that, when a electron cloud radiates coherently, the radiation losses per electron are much larger than those for incoherent radiation. Therefore, for a sufficiently large number of electrons, the coherent radiation friction must be taken into account from the very beginning.

In its general formulation, this problem has not been solved yet (see, e.g., [5.9]); we will address it in detail elsewhere. To account for the coherent radiation in this Letter, we approximate a small and dense electron cloud in a strong laser field by a *single pointlike particle* with the charge $q = N_e e$ and mass $m = N_e m_e$, which we call a "fat electron." This model is appropriate because of the strictly field-driven nature of the cloud motion, whereby the tight cloud can be ascribed a single trajectory identical to trajectories of individuals electrons. As discussed above, we neglect the electron motion along the direction of laser propagation; the equation of motion in the plane normal to that axis is then

$$\frac{d\vec{p}}{dt} + \Gamma_{fat}\omega_L \vec{p} \approx e\vec{E}_L, \qquad \Gamma_{fat} \equiv N_e \Gamma_e \gamma^3, \tag{5.7}$$

where Γ_{fat} and $\Gamma_{fat}\omega_L \vec{p}$ are the radiation damping constant and the "radiation friction" force of the fat electron, respectively (for a single electron, see, e.g., [5.6]). At the equilibrium between the friction and driving forces, the relativistic factor, instead of (5.2), becomes now

$$\gamma = \sqrt{1 + \frac{\varepsilon_L^2}{1 + \Gamma_{fat}^2}}. \tag{5.8}$$

For a *single* electron, we have $\Gamma_{fat} << 1$, and Eq. (5.2) is accurate for laser power up td thousands of petawatts. As N_e increases, however, Γ_{fat} grows proportionally. One can understand this also in terms of the ratio of energy radiated by a fat electron per radian of rotation angle, $W_{rad} = N_e^2 \Gamma_{fat} m_e c^2 \gamma^2 (\gamma^2 - 1)$, to its kinetic energy $W_{rot} = N_e m_e c^2 (\gamma^2 - 1)$:

$$\Gamma_{fat} = \frac{1}{1 - \gamma^{-1}} \frac{W_{rad}}{W_{rot}}, \tag{5.9}$$

hence, the increase in Γ_{fat} signifies the increase of radiation energy compared to that of the driven rotation; at $\gamma >> 1$, $\Gamma_{fat} \approx W_{rad} / W_{rot}$. Introducing an "EM size" of the fat electron as $r_{fat} = r_e N_e \gamma^3$, we see that $\Gamma_{fat} = 1$ when $r_{fat} = (3/4\pi)\lambda_L$. The conventional theory of electron radiation at $\hbar\omega << m_e c^2$ is based on the fact that $r_e <<< \lambda_L$; i.e., the radiation energy is very small compared to a kinetic one; it breaks down completely if $r_{fat} = O(\lambda)$. In the weak radiation case, a strong driving field, $\varepsilon_L^2 >> 1$ would result in $\gamma \approx \varepsilon_L$. However, if the radiation is strong and coherent, the further increase of electron energy γ is drastically inhibited as ε_L^2 increases:

$$\gamma \approx \left(\frac{\varepsilon_L}{\Gamma_N}\right)^{\frac{1}{4}} \text{ if } \varepsilon_L^2 >> \Gamma_N^2 + \Gamma_N^{-2/3}, \tag{5.10}$$

where $\Gamma_N = N_e\Gamma_e$. The strong radiation friction results in a trade-off between the pulse duration and radiated energy: More energy requires more electrons, which in turn limits γ and τ_{pl}. This still allows for spectacular output. For example, if N_e is such that, for $\varepsilon_L = 100$ (PL), Eq. (5.8) yields $\gamma \approx (2/3)\varepsilon_L$, then $N_e \approx 300$, and one may expect EM bursts of 0.9 zs, separated by 3 fs intervals, each burst carrying 3 fJ energy with the spectral cutoff at 1.2 MeV. If $N_e = 21000$, the energy/burst grows to 5 pJ, but γ drops to $(1/4)\varepsilon_L$, so that $\tau_{pl} \sim 17$ zs is still very short.

This trade-off can be circumvented by combining the lasetron with a relativistic heavy ion accelerator. If the lasetron pulses are directed toward a uranium ion beam with $\gamma_{nucl} \approx 100$, as in RHIC at BNL, then uranium nuclei would see in their rest frame the Doppler up-shifted pulses of $\tau_{pl}/2\gamma_{nucl}$ duration. For the second example above, the up-shifted spectrum is concentrated in the area of uranium giant dipole resonance (the maximum of photofission cross section), while the burst duration is short compared with the lifetime of fissioning nuclei. Therefore, a combination of the petawatt laser driven lasetron and RHIC holds a potential for time-resolved measurements and control of fast nuclear fission. Moreover, the nuclei accelerated in the future Large Hadron Collider ($\gamma_{nucl} \approx 3000$) would see lasetron pulses shortened to yet another, yoctosecond (10^{-24} s) domain.

In the paper [5.10] the theoretical project of the *LASETRON* was described. As was shown in paper [5.10] the 10^{-21} s (zeptosecond) laser pulses can be generated using petawatt lasers, while already available terawatt lasers may generate subattosecond laser pulses of 10^{-19} s. The pulses will be radiated by ultrarelativistic electrons driven by circularly polarized high-intensity laser fields. *LASETRON* pulses can be achieved by placing a solid particle or a piece of wire of subwavelength cross section in the focal plane of a superpowerful laser.

In the book [5.11] it was shown that the lifetime of the electron-positron pair is of the order of 10^{17}s. Strictly speaking the lifetime of the order of the relaxation time τ

$$\tau = \frac{\hbar}{m_e \alpha^2 c^2} \approx 10^{-17}\ \text{s}. \tag{5.11}$$

In formula (5.11) $m_e = m_p$ is electron, positron mass, α is the fine structure constant and c is the vacuum light velocity. As can be concluded from formula (5.11) the *zitterbegung* or *tremor* of space-time can be investigated with the *LASETRON* pulses, for the latter are shorter than the characteristic relaxation time.

For the time period $\Delta t < \tau$, i.e. for *LASETRON* pulse the vacuum of space-time is filled with the gas of electron-positron pairs with lifetime $\approx 10^{-17}$ s. In that case the propagation of *LASETRON* pulse will be described by the Heaviside equation:

$$\frac{\partial^2 E}{\partial t^2} + \frac{\sigma}{\varepsilon_0}\frac{\partial E}{\partial t} = c_\gamma^2 \frac{\partial^2 E}{\partial x^2}. \tag{5.12}$$

In equation (5.11) ε_0 - permittivity of free space-time, and σ is the conductivity of free space-time, c_γ is the photon velocity in space-time and E is electric field (in one dimension).

In the seminal paper [5.12] F. Calogero described the cosmic origin of quantization. In paper [5.12] the *tremor* of the cosmic particles is the origin of the quantization and the characteristic acceleration of these particles $a \approx 10$ m/s^2 was calculated. In our earlier paper [5.13] the same value of the acceleration was obtained and compared to the experimental value of the measured space-time acceleration [5.14]. In this paper we define the cosmic force — *Planck* force, $F_{Planck} = M_P\, a_{Planck}$ ($a_{Planck} \approx a$) and study the history of Planck force as the function of the age of the Universe.

Masses introduce a curvature in space-time, light and matter are forced to move according to space-time metric. Since all the matter is in motion, the geometry of space is constantly changing. A. Einstein relates the curvature of space to the mass/energy density:

$$\mathbf{G} = k\,\mathbf{T}, \tag{5.13}$$

G is the Einstein curvature tensor and **T** the stress-energy tensor. The proportionality factor k follows by comparison with Newton's theory of gravity: $k = G/c^4$, where G is the Newton's gravity constant and c is the vacuum velocity of light; it amounts to about 2.10^{-43} N^{-1} expressing the *rigidity* of space-time.

In paper [5.13] the model for the acceleration of space-time was developed. Prescribing the $-G$ for space-time and $+G$ for matter the acceleration of space-time was obtained:

$$a_{Planck} = -\frac{1}{2}\left(\frac{\pi}{4}\right)^{\frac{1}{2}} \frac{\left(N+\frac{3}{4}\right)^{\frac{1}{2}}}{M^{\frac{3}{2}}} A_P, \tag{5.14}$$

where A_P, *Planck* acceleration equal, viz.;

$$A_P = \left(\frac{c^7}{\hbar G}\right)^{\frac{1}{2}} = \frac{c}{\tau_P} \cong 10^{51}\,\mathrm{ms}^{-2}. \tag{5.15}$$

As was shown in paper [5.13] the a_{Planck} for $N = M = 10^{60}$ is of the order of the acceleration detected by Pioneer spacecrafts [5.14].

Considering A_P it is quite natural to define the *Planck* force F_{Planck}

$$F_{Planck} = M_P A_P = \frac{c^4}{G} = k^{-1}, \tag{5.16}$$

where

$$M_P = \left(\frac{\hbar c}{G}\right)^{\frac{1}{2}}.$$

From formula (5,16) we conclude that $\left(F_{Planck}\right)^{-1}$ = rigidity of the space-time. The *Planck* force, $F_{Planck} = c^4/G = 1.2\cdot10^{44}$ N can be written in units which characterize the microspace-time, i.e. GeV and fm. In that units

$k^{-1} = F_{Planck} = 7.6\cdot10^{38}$ GeV/fm.

As was shown in paper [5.13] the present value of *Planck* force equal

$$F_{Planck}^{Now}(N = m = 10^{60}) \cong -\frac{1}{2}\left(\frac{\pi}{4}\right)^{\frac{1}{2}} 10^{-60}\,\frac{c^4}{G} = -10^{-22}\,\frac{\text{GeV}}{\text{fm}}. \tag{5.17}$$

In papers [5.15, 5.16] the *Planck* time τ_P was defined as the relaxation time for space-time

$$\tau_P = \frac{\hbar}{M_P c^2}. \tag{5.18}$$

Considering formulae (5.16) and (5.18) F_{Planck} can be written as

$$F_{Planck} = \frac{M_P c}{\tau_P}, \tag{5.19}$$

where c is the velocity for gravitation propagation. In papers [5.15, 5.16] the velocities and relaxation times for thermal energy propagation in atomic and nuclear matter were calculated:

$$\begin{aligned} \upsilon_{atomic} &= \alpha_{em} c, \\ \upsilon_{nuclear} &= \alpha_s c, \end{aligned} \tag{5.20}$$

where $\alpha_{em} = e^2/(\hbar c) = 1/137,\ \alpha_s = 0.15$. In the subsequent we define atomic and nuclear accelerations:

$$a_{atomic} = \frac{\alpha_{em} c}{\tau_{atomic}},$$
$$a_{nuclear} = \frac{\alpha_s c}{\tau_{nuclear}}. \tag{5.21}$$

Considering that $\tau_{atomic} = \hbar/\left(m_e \alpha_{em}^2 c^2\right)$, $\tau_{nuclear} = \hbar/\left(m_N \alpha_s^2 c^2\right)$ one obtains from formula (5.21)

$$a_{atomic} = \frac{m_e c^3 \alpha_{em}^3}{\hbar},$$
$$a_{nuclear} = \frac{m_N c^3 \alpha_s^3}{\hbar}. \tag{5.22}$$

We define, analogously to *Planck* force the new forces: F_{Bohr}, F_{Yukawa}

$$F_{Bohr} = m_e a_{atomic} = \frac{\left(m_e c^2\right)^2}{\hbar c} \alpha_{em}^3 = 5 \cdot 10^{-13} \frac{\text{GeV}}{\text{fm}},$$
$$F_{Yukawa} = m_N a_{nuclear} = \frac{\left(m_N c^2\right)^2}{\hbar c} \alpha_s^3 = 1.6 \cdot 10^{-2} \frac{\text{GeV}}{\text{fm}}. \tag{5.23}$$

Comparing formulae (5.17) and (5.23) we conclude that gradients of *Bohr* and *Yukawa* forces are much large than F_{Planck}^{Now}, i.e.:

$$\frac{F_{Bohr}}{F_{Planck}^{Now}} = \frac{5 \cdot 10^{-13}}{10^{-22}} \cong 10^9,$$
$$\frac{F_{Yukawa}}{F_{Planck}^{Now}} = \frac{10^{-2}}{10^{-22}} \cong 10^{20}. \tag{5.24}$$

The formulae (5.24) guarantee present day stability of matter on the nuclear and atomic levels.

As the time dependence of F_{Bohr} and F_{Yukawa} are not well established, in the subsequent we will assumed that α_s and α_{em} [5.17] do not dependent on time. Considering formulae (5.19) and (5.22) we obtain

$$\frac{F_{Yukawa}}{F_{Planck}} = \frac{1}{\left(\frac{\pi}{4}\right)^{\frac{1}{2}}} \frac{\left(m_N c^2\right)^2}{M_P c^2} \frac{\alpha_s^3}{\hbar} T, \tag{5.25}$$

$$\frac{F_{Bohr}}{F_{Planck}} = \frac{1}{\left(\frac{\pi}{4}\right)^{\frac{1}{2}}} \frac{\left(m_e c^2\right)^2}{M_P c^2} \frac{\alpha_{em}^3}{\hbar} T. \tag{5.26}$$

As can be realized from formulae (5.25), (5.26) in the past $F_{Planck} \approx F_{Yukawa}$ (for $T = 0.002$ s) and $F_{Planck} \approx F_{Bohr}$ (for $T \approx 10$ s), T = age of universe.The calculated ages define the limits for instability of the nuclei and atoms.

In 1900 M. Planck [5.18] introduced the notion of the universal mass, later on called the *Planck* mass

$$M_P = \left(\frac{\hbar c}{G}\right)^{\frac{1}{2}}. \tag{5.27}$$

Considering the definition of the *Yukawa* force (5.23)

$$F_{Yukawa} = \frac{m_N \upsilon_N}{\tau_N} = \frac{m_N \alpha_{strong} c}{\tau_N}, \tag{5.28}$$

the formula (5.28) can be written as:

$$F_{Yukawa} = \frac{m_{Yukawa} c}{\tau_N}, \tag{5.29}$$

where

$$m_{Yukawa} = m_N \alpha_{strong} \cong 147 \frac{\text{MeV}}{c^2} \sim m_\pi. \tag{5.30}$$

From the definition of the *Yukawa* force we deduced the mass of the particle which mediates the strong interaction – pion mass postulated by Yukawa in [5.19].

Accordingly for *Bohr* force:

$$F_{Bohr} = \frac{m\,\upsilon}{\tau_{Bohr}} = \frac{m_e \alpha_{em} c}{\tau_{Bohr}} = \frac{m_{Bohr} c}{\tau_{Bohr}}, \tag{5.31}$$

$$m_{Bohr} = m_e \alpha_{em} = 3.7 \frac{\text{keV}}{c^2}. \tag{5.32}$$

For the *Bohr* particle the range of interaction is

$$\gamma_{Bohr} = \frac{\hbar}{m_{Bohr}c} \approx 0.1\,\text{nm}, \tag{5.33}$$

which is of the order of atomic radius.

Considering the electromagnetic origin of the mass of the *Bohr* particle, the planned sources of hard electromagnetic field *LASETRON* [5.10] are best suited to the investigation of the properties of the *Bohr* particles.

In an important work, published already in 1951 J. Schwinger [5.20] demonstrated that in the background of a static uniform electric field, the QED space-time is unstable and decayed with spontaneous emission of $e^+ e^-$ pairs. In the paper [5.20] Schwinger calculated the critical field strengths E_S:

$$E_S = \frac{m_e^2 c^3}{e\hbar}. \tag{5.34}$$

Considering formula (5.34) we define the *Schwinger* force:

$$F^e_{Schwinger} = eE_S = \frac{m_e^2 c^3}{\hbar}. \tag{5.35}$$

Formula (5.35) can be written as:

$$F^e_{Schwinger} = \frac{m_e c}{\tau_{Sch}}, \tag{5.36}$$

where

$$\tau_{Sch} = \frac{\hbar}{m_e c^2} \tag{5.37}$$

is *Schwinger* relaxation time for the creation of $e^+ e^-$ pair. Considering formulae (5.23) the relation of F_{Yukawa} and F_{Bohr} to the *Schwinger* force can be established

$$F_{Yukawa} = \alpha_s^3 \left(\frac{m_N}{m_e}\right)^2 F^e_{Schwinger}, \alpha_s = 0.15,$$
$$F_{Bohr} = \alpha_{em}^3 F^e_{Schwinger}, \alpha_{em} = \frac{1}{137}, \tag{5.38}$$

and for *Planck* force

$$F_{Planck} = \left(\frac{M_P}{m_e} \right)^2 F^e_{Schwinger}. \tag{5.39}$$

In Table 5.2 the values of the $F^e_{Schwinger}$, F_{Planck}, F_{Yukawa} and F_{Bohr} are presented, all in the same units GeV/fm. As in those units the forces span the range 10^{-13} to 10^{38} it is valuable to recalculate the *Yukawa* and *Bohr* forces in the units natural to nuclear and atomic level. In that case one obtains:

$$F_{Yukawa} = 16 \frac{\text{MeV}}{\text{fm}}. \tag{5.40}$$

It is quite interesting that $\alpha_v \approx 16$ MeV is the volume part of the binding energy of the nuclei (droplet model).

Table 5.2. Schwinger, Planck, Yukawa and Bohr forces [GeV/fm]

$F^e_{Schwinger}$	FPlanck	FYukawa	FBohr
≈ 10-6	≈ 1038	≈ 10-2	≈ 10-13

For the *Bohr* force considering formula (5.23) one obtains:

$$F_{Bohr} = \frac{50\,\text{eV}}{0.1\,\text{nm}}. \tag{5.41}$$

Considering that the *Rydberg* energy ≈ 27 eV and *Bohr* radius ≈ 0.1 nm formula (5.41) can be written as

$$F_{Bohr} = \frac{\text{Rydberg energy}}{\text{Bohr radius}}. \tag{5.42}$$

5.2. Fundamental Physics at an X-Ray Free Electron Laser

There are definite plans for the construction of free electron lasers (FELs) in the X-ray band, both at the Standord Linear Accelerator Center (SLAC), where the so-called Linac Coherent Light Source (LCLS) has been proposed [5.21] as well as at DESY, where the so-called XFEL laboratory is part of the design of the electron-positron ($e^+ e^-$) linear collider TESLA (TeV-Energy Superconducting Linear Accelerator) [5.22, 5.23].

X-ray free electron lasers will give us new insights into natural and life sciences. X-rays play a crucial role when the structural and electronic properties of matter are to be studied on an atomic scale. The spectral characteristics of the planned X-ray FELs, with their high power, short pulse length, narrow bandwidth, spatial coherence, and a tunable wavelength, make them ideally suited for applications in atomic and molecular physics, plasma physics, condensed matter physics, material science, chemistry, and structural biology [5.21 – 5.23].

In addition to these immediate applications, X-ray FELs may be employed also to study some physics issues of fundamental nature [5.24]. In this context, one may mention the boiling of the vacuum [5.25 – 5.29] (Schwinger pair creation in an external field), horizon physics [5.30, 5.31] (Unruh effect), and axion production [5.32, 5.33].

Conventional lasers yield radiation typically in the optical band. The reason is that in these devices the gain comes from stimulated emission from electrons bound to atoms, either in a crystal, liquid dye, or a gas. The amplification medium of free electron lasers [5.34], on the other hand, is *free* (unbounded) electrons in bunches accelerated to relativistic velocities with a characteristic longitudinal charge density modulation. The radiation emitted by an FEL can be tuned over a wide range of wavelengths, which is a very important advantage over conventional lasers.

The basic principle of a single-pass free electron laser operating in the self amplified spontaneous emission (SASE) mode [5.35] is as follows. It functions by passing an electron beam pulse of energy E_e of small cross section and high peak current through a long periodic magnetic structure (undulator). The interaction of the emitted synchrotron radiation, with opening angle $1/\gamma$

$$\frac{1}{\gamma}=\frac{m_e c^2}{E_e}=2\cdot 10^{-5}(25\,\mathrm{GeV}/E_e), \tag{5.43}$$

where m_e is the electron mass, with the electron beam pulse within the undulator leads to the buildup of a longitudinal charge density modulation (micro bundling), if a resonance condition,

$$\lambda=\frac{\lambda_U}{2\gamma^2}\left(1+\frac{K_U^2}{2}\right)=0.3\,\mathrm{nm}\left(\frac{\lambda_U}{1\,\mathrm{m}}\right)\left(\frac{1/\gamma}{2\cdot 10^{-5}}\right)^2\left(\frac{1+K_U^2/2}{3/2}\right), \tag{5.44}$$

is met. Here, λ is the wavelength of the emitted radiation, λ_U is the length of the magnetic period of the undulator, and K_U is the undulator parameter,

$$K_U=\frac{e\lambda_U B_U}{2\pi m_e c}, \tag{5.45}$$

which gives the ratio between the average deflection angle of the electrons in the undulator magnetic field B_U from the forward direction and the typical opening cone of the synchrotron radiation. The undulator parameter should be of order one on resonance. The electrons in the developing micro bunches eventually radiate coherently - the gain in radiation power P,

$$P\propto e^2 N_e^2 B_U^2 \gamma^2, \tag{5.46}$$

over the one from incoherent spontaneous synchrotron radiation ($P \propto N_e$) being proportional to the number $N_e \geq 10^9$ of electrons in a bunch – and the number of emitted photons grows exponentially until saturation is reached. The radiation has a high power, short pulse length, narrow bandwidth, is folly polarized, transversely coherent, and has a tunable wavelength.

The concept of using a high energy electron linear accelerator for building an X-ray FEL was first proposed for the Stanford Linear Accelerator [5.21]. The feasability of a single-pass FEL operating in the S ASE mode has recently been demonstrated [5.36] down to a wavelength of 80 nm using electron bundles of high charge density and low emittance from the linear accelerator at the TESLA test facility (TTF) at DESY. An X-ray FEL laboratory is planned as an integral part of TESLA [5.22, 5.23]. Some characteristics of the radiation from the planned X-ray FELs at TESLA are listed in Table 5.3.

Table 5.3. Typical photon beam properties of the SASE FELs at TESLA [5.23].

unit	SASE 1	SASE 2	SASE 3	SASE 4	SASE 5
Wavelength [nm]	0.1 – 0.5	0.085 – 0.133	0.1 – 0.24	0.1 – 1	0.4 – 5.8
Peak power [GW]	37	19	22	30	110 – 200
Average power [W]	210	110	125	170	610 – 1100
Number photons per pulse	$1.8 \cdot 10^{12}$	$8.2 \cdot 10^{11}$	$1.1 \cdot 10^{12}$	$1.5 \cdot 10^{12}$	$2.2 - 58 \cdot 10^{13}$
Bandwidth (FWHM) [%]	0.08	0.07	0.08	0.08	0.29 – 0.7
Pulse duration (FWHM) [fs]	100	100	100	100	100

The spectral characteristics of X-ray free electron lasers suggest immediate applications in condensed matter physics, chemistry, material science, and structural biology, which are reviewed in the conceptual [5.21, 5.22] and technical [5.23] design reports of the planned X-ray FEL facilities. In this section, we want to emphasize that X-ray FELs may be employed also to study some particle physics issues. Whereas the last application mainly requires large average radiation power $\langle P \rangle$, the first two applications require very large electric fields and thus high peak power densities $P/(\pi\sigma^2)$ where σ is the laser spot radius. Here, one could make use of the possibility to focus the X-ray beam to a spot of small radius, hopefully down to the diffraction limit, $\sigma \geq \lambda \cong O(0.1)$ nm. In this way, one may obtain very large electric fields and accelerations,

$$\varepsilon = \sqrt{\mu_0 c \frac{P}{\pi\sigma^2}} = 1.1 \cdot 10^{17} \, \frac{\mathrm{V}}{\mathrm{m}} \left(\frac{P}{1\,\mathrm{TW}} \right)^{\frac{1}{2}} \left(\frac{0.1\,\mathrm{nm}}{\sigma} \right), \tag{5.47}$$

$$a = \frac{e\varepsilon}{m_e} = 1.9 \cdot 10^{28} \frac{\text{m}}{\text{s}^2} \left(\frac{P}{1\,\text{TW}} \right)^{\frac{1}{2}} \left(\frac{0.1\,\text{nm}}{\sigma} \right), \tag{5.48}$$

much larger than those obtainable with any optical laser of the same power.

Spontaneous particle creation from vacuum induced by an external field, first put forth to examine the production of e^+e^- pairs in a static, spatially uniform electric field [5.37] and often referred to as the Schwinger mechanism, ranks among the most intriguing nonlinear phenomena in quantum field theory. Its consideration is theoretically important, since it requires one to go beyond perturbation theory, and its experimental observation would verify the validity of the theory in the domain of strong fields. Moreover, this mechanism has been applied to many problems in contemporary physics, ranging from black hole quantum evaporation [5.38] to particle production in hadronic collisions [5.39] and in the early universe [5.40], to mention only a few. One may consult the monographs [5.41] for a review of further applications, concrete calculations and a detailed bibliography.

It is known since a long time that in the background of a static, spatially uniform electric field the vacuum in quantum electrodynamics (QED) is unstable and, in principle, sparks with spontaneous emission of e^+e^- pairs [5.37]. However, a sizeable rate for spontaneous pair production requires extraordinary strong electric field strengths ε of order or above the critical value

$$\varepsilon_c \equiv \frac{m_e c^2}{e \lambda\!\!\!^{-}_e} = \frac{m_e^2 c^3}{e\hbar} \cong 1.3 \cdot 10^{18}\ \text{V/m}. \tag{5.49}$$

Otherwise, for $\varepsilon << \varepsilon_c$, the work of the field on a unit charge e over the Compton wavelength of the electron $\lambda\!\!\!^{-}_e = \hbar/(m_e c)$ is much smaller than the rest energy $2m_e c^2$ of the produced e^+e^- pair, the process can occur only via quantum tunneling, and its rate is exponentially suppressed,

$$\frac{\text{d}^4 n_{e^+e^-}}{\text{d}^3 x \text{d}t} \sim \frac{c}{4\pi^3 \lambda\!\!\!^{-}_e{}^4} \exp\left[-\pi \frac{\varepsilon_c}{\varepsilon} \right]. \tag{5.50}$$

Unfortunately, it seems inconceivable to produce macroscopic static fields with electric field strengths of the order of the Schwinger critical field (5.49) in the laboratory. In view of this difficulty, in the early 1970's the question was raised whether intense optical lasers could be employed to study the Schwinger mechanism [5.42, 5.43]. Yet, it was found that all available and conceivable optical lasers did not have enough power density to allow for a sizeable pair creation [5.42 – 5.52].

With the possibility of X-ray lasers at the horizon, this question has been addressed recently again [5.25 – 5.29]. As a quasi-realistic picture of the electromagnetic field of a laser, a pure electric field oscillating with a frequency $\omega = 2\pi c/\lambda$ was considered, under the assumption that the field amplitude ε is much smaller than the Schwinger critical field, and the photon energy is much smaller than the rest energy of the electron,

$$\varepsilon << \varepsilon_c, \qquad \hbar\omega << m_e c^2; \tag{5.51}$$

conditions which are well satisfied at realistic X-ray lasers. Under these conditions, it is possible to compute the rate of e^+e^- pair production in a semiclassical manner, using generalized WKB or imaginary-time methods [5.43, 5.44, 5.46, 5.48, 5.51, 5.52]. Here, the ratio η of the energy of the laser photons over the work of the field on a unit charge e over the Compton wavelength of the electron,

$$\eta = \frac{\hbar\omega}{e\varepsilon\lambda\!\!\!^{-}_e} = \frac{\hbar\omega}{m_e c^2}\frac{\varepsilon_c}{\varepsilon} = \frac{m_e c\omega}{e\varepsilon}, \tag{5.52}$$

plays the role of an adiabaticity parameter. As long as $\eta << 1$, i. e. in the strong-field, low-frequency limit, the non-perturbative Schwinger result (5.50) for a static uniform field applies. On the other hand, for large η, i. e. in the low-field, high-frequency limit, the essentially perturbative result

$$\frac{\mathrm{d}^4 n_{e^+e^-}}{\mathrm{d}^3x\mathrm{d}t} \sim \frac{c}{4\pi^3\lambda\!\!\!^{-}{}^4_e}\left(\frac{\mathrm{e}}{4}\frac{eE\lambda\!\!\!^{-}_e}{\hbar\omega}\right)^{\frac{4m_ec^2}{\hbar\omega}} \tag{5.53}$$

is obtained for the rate of pair production. It corresponds to the n-th order perturbation theory, n being the minimum number of quanta required to create an e^+e^- pair: $n \geq 2m_e c^2/(\hbar\omega) >> 1$.

For an X-ray laser, with $\hbar\omega \sim 1-10$ keV, the adiabatic, nonperturbative, strong field regime, $n \leq 1$, starts to apply for $\varepsilon \geq \hbar\omega\varepsilon_c/(m_e c^2) \sim 10^{15\div16}$ V/m (c.f. Eq. (5.52)). An inspection of the tunneling rate (5.50) leads then to the conclusion that one needs an electric field of about $0.1\varepsilon_c \sim 10^{17}$ V/m in order to get an appreciable amount of spontaneously produced e^+e^- pairs at an X-ray laser [5.27]. Under such conditions the production rate is time-dependent, with repeated cycles of particle production and annihilation in tune with the laser frequency, but the peak particle number is independent of the laser frequency: up to 10^3 pairs may be produced in the spot volume [5.28].

Time-resolved experiments are used to monitor time-dependent phenomena. The study of dynamics in physical systems often requires time resolution beyond the femtosecond capabilities. Subfemtosecond capabilities are now available in the XUV wavelength range [5.53, 5.54]. This is achieved by focusing a fs laser into a gas target creating radiation of high harmonics of fundamental laser frequency. In principle, table-top ultra-fast X-ray sources have the right duration to provide us with a view of subatomic transformation processes. However, their power and photon energy are by far low. There also exists a wide interest in the extension of attosecond techniques into the 0.1 nm wavelength range.

With the realization of the fourth-generation light sources operating in the X-ray regime [5.55, 5.56], new attoscience experiments will become possible. In its initial configuration the XFEL pulse duration is about 100 fs, which is too long to be sufficient for this class of experiments. The generation of subfemtosecond X-ray pulses is critical to exploring the ultrafast science at the XFELs. The advent of attosecond X-ray pulses will open a new field of time-resolved studies with unprecedented resolution. X-ray SASE FEL holds a great promise as a source of radiation for generating high power, single attosecond pulses. Recently a scheme to achieve pulse duration down to attosecond time scale at the wavelengths around 0.1 nm has been proposed [5.57]. It has been shown that by using X-ray SASE FEL combined with terawatt-level, sub-10-fs Ti:sapphire laser system it will be possible to produce GW-level X-ray pulses that are reaching 300 attoseconds in duration. In this scheme an ultrashort laser pulse is used to modulate the energy of electrons within the femtosecond slice of the electron bunch at laser frequency. Energy-position correlation in the electron pulse results in spectrum-position correlation in the SASE radiation pulse. Selection of ultra-short X-ray pulses is achieved by using the monochromator. Such a scheme for production of single attosecond X-ray pulses would offer the possibility for pump-probe experiments, since it provides a precise, known and tunable interval between the laser and X-ray sources.

In the paper [5.58] the authors propose a new method allowing to increase output power of attosecond X-ray pulses by two orders of magnitude. It is based on application of sub-10-fs laser for slice energy modulation of the electron beam, and application "fresh bunch" techniques for selection of single attosecond pulses with 100 GW-level output power. The combination of very high peak power (100 GW) and very short pulse (300 as) will open a vast new range of applications. In particular, we propose visible pump/X-ray probe technique that would allow time resolution down to subfemtosecond capabilities. Proposed technique allows to produce intense ultrashort X-ray pulses directly from the XFEL, and with tight synchronization to the sample excitation laser. Another advantage of the proposed scheme is the possibility to remove the monochromator (and other X-ray optical elements) between the X-ray undulator and a sample and thus to directly use the probe attosecond X-ray pulse.

An ultrashort laser pulse is used to modulate the energy of electrons within the femtosecond slice of the electron bunch at laser frequency. The seed laser pulse will be timed to overlap with the central area of the electron bunch. It serves as a seed for modulator which consists of a short (a few periods) undulator. Following the energy modulator the beam enters the baseline (gap-tunable) X-ray undulator. In its simplest configuration the X-ray undulator consists of an uniform input undulator and nonuniform (tapered) output undulator separated by a magnetic chicane (delay). The process of amplification of radiation in the input undulator develops in the same way as in conventional X-ray SASE FEL: fluctuations of the electron beam current serve as the input signal [5.59]. When an electron beam traverses ah undulator, it emits radiation at the resonance wavelength $\lambda = \lambda_w \left(1 + K^2/2\right)/\left(2\gamma^2\right)$. Here λ_w is the undulator period, $mc^2\gamma$ is the electron beam energy, and K is the undulator parameter. In the proposed scheme the laser-driven sinusoidal energy chirp produces a correlated frequency chirp of the resonant radiation $\delta\omega/\omega \cong 2\delta\gamma/\gamma$.

Our concept of attosecond X-ray facility is based on the use of a few cycle optical pulse from Ti:sapphire laser system. This optical pulse is used for modulation of the energy of the electrons within a slice of the electron bunch at a wavelength of 800 nm. Due to extreme temporal confinement, moderate optical pulse energies of the order of a few mJ can result in electron energy modulation amplitude higher than 30-40 MeV. In few-cycle laser fields high intensities can be "switched on" nonadiabatically within a few optical periods. As a result, a central peak electron energy modulation is larger than other peaks. This relative energy difference is used for selection of SASE radiation pulses with a single spike in time domain. Single-spike selection can effectively be achieved when electron bunch passes through a magnetic delay and output undulator operating at a shifted frequency.

5.3. Π – Mesons as the Quanta of Non-Newtonian Hadronic Fluid

As is well known Nelson [5.60] in 1966 succeed in deriving the Schrödinger equation from the assumption that quantum particles follow continuous trajectories in a chaotic background. The derivation of the usual linear Schrödinger equation follows only if the diffusion coefficient D, associated with quantum brownian motion takes the value $D = \hbar/2m$ as assumed by Nelson.

In this chapter we study the transfer process of the quantum particles in the context of the thermal energy transport in highly excited matter. It will be shown that when matter is excited with short thermal perturbation the response of the matter can be well described by quantum hyperbolic heat transfer equation (QHT) which is the generalization of the parabolic quantum heat transport equation (PHT) with the diffusion coefficient $D = \hbar/2m$, where m denotes the mass of the diffused particles. The obtained QHT has the form

$$\frac{1}{c^2}\frac{\partial^2 T}{\partial t^2} + \frac{1}{c^2\tau}\frac{\partial T}{\partial t} = \frac{\alpha^2}{3}\nabla^2 T. \tag{5.54}$$

where T denotes temperature, c is the velocity of light and α_i = (1/137, 0.15, 1) is the fine structure constant for electromagnetic interaction, strong interaction and strong quark-quark interaction respectively, τ_i is the relaxation time for scattering process.

When the QHT is applied to the study of the thermal excitation of the matter, the quanta of thermal energy, the heaton can be defined with energies $E_h^e = 9$ eV, $E_h^H = 7$ MeV and $E_h^q = 139$ MeV for atomic, nucleon and quark level respectively.

One of the best models in mathematical physics is Fourier's model for the heat conduction in matter. Despite the excellent agreement obtained between theory and experiment, the Fourier model contains several inconsistent implications. The most important is that the model implies an infinite speed of propagation for heat. Cattaneo [5.61] was the first to propose a remedy. He formulated new hyperbolic heat diffusion equation for propagation of the heat waves with finite velocity.

There is an impressive amount of literature on hyperbolic heat trans-port in matter [5.62 – 5.64]. In our book [5.11] we developed the new hyperbolic heat transport equation which generalizes the Fourier heat transport equation for the rapid thermal processes. The hyperbolic heat conduction equation (HHC) for the fermionic system can be written in the form:

$$\frac{1}{\left(\frac{1}{3}\upsilon_F^2\right)}\frac{\partial^2 T}{\partial t^2}+\frac{1}{\tau\left(\frac{1}{3}\upsilon_F^2\right)}\frac{\partial T}{\partial t}=\nabla^2 T, \tag{5.55}$$

where T denotes the temperature, τ – the relaxation time for the thermal disturbance of the fermionic system and υ_F is the Fermi velocity.

When the ordinary matter (on the atomic level) or nuclear matter (on the nucleus level) is excited with short temperature pulses ($\Delta t \sim \tau$) the response of the matter is discrete. The matter absorbs the thermal energy in the form of the quanta E_h^e or E_h^N.

It is quite natural to pursue the study of the thermal excitation to the subnucleon level i.e. quark matter. In the following we generalize the QHT equation for quark gas in the form:

$$\frac{1}{c^2}\frac{\partial^2 T^q}{\partial t^2}+\frac{1}{c^2\tau}\frac{\partial T^q}{\partial t}=\frac{\left(\alpha_s^q\right)^2}{3}\nabla^2 T^q \tag{5.56}$$

with α_s^q – the fine structure constant for strong quark-quark interaction and υ_s^q - thermal velocity:

$$\upsilon_h^q=\frac{1}{\sqrt{3}}\alpha_s^q c. \tag{5.57}$$

Analogously as for electron and nucleon gases we obtain for quark heaton

$$E_h^q=\frac{m}{3}\left(\alpha_s^q\right)^2 c^2, \tag{5.58}$$

where m_q denotes the mass of the average quark mass. For quark gas the average quark mass can be calculated according to formula [5.65]

$$m_q=\frac{m_u+m_d+m_s}{3}=\frac{350\,\text{MeV}+350\,\text{MeV}+550\,\text{MeV}}{3}=417\,\text{MeV}, \tag{5.59}$$

where m_u, m_d, m_s denotes the mass of the up, down and strange quark respectively. For the calculation of the α_s^q we consider the decays of the baryon resonances. For strong decay of the Σ^0(1385 MeV) resonance:

$$K^- + p \to \Sigma^0(1385\,\text{MeV}) \to \Lambda + \pi^0$$

width $\Gamma \sim 36$ MeV and lifetime τ_S.

For electromagnetic decay

$$\Sigma^0(1192\,\text{MeV}) \to \Lambda + \gamma,$$

$\tau_e \sim 10^{-19}$ s. Considering that

$$\left(\frac{\alpha_s^q}{\alpha}\right) \sim \left(\frac{\tau_e}{\tau_s}\right)^{\frac{1}{2}} \sim 100,$$

one obtains for α_s^q the value

$$\alpha_s^q \sim 1. \tag{5.60}$$

Substituting formulae (5.59), (5.60) to formula (5.58) one obtains:

$$E_h^q \sim 139\,\text{MeV} \sim m_\pi, \tag{5.61}$$

where m_π denotes the π – meson mass. It occurs that when we attempt to "melt" the nucleon in order to obtain the free quark gas the energy of the *heaton* is equal to the π – meson mass (which consists of two quarks). It is the simple presentation of quark confinement. Moreover it seems that the standard approaches to the melting of the nucleons into quarks through the heating processes in "splashes" of the chunks of the nuclear matter do not promise the success.

We argue that at excitation energy of the order of pion mass the hadronic matter undergoes the phase transition to the non-Newtonian fluid with the relaxation time, τ

$$\tau = \frac{\hbar}{m_q c^2}, \tag{5.62}$$

where $m_q c^2$ is of the order of 400 MeV and the velocity of sound $v_h \sim c$. The thermal energy quanta in non-Newtonian hadronic fluid are of the order of 140 MeV $\cong m_\pi$. In this model the π – meson can be described as the "phonon" – excitations of the non-Newtonian hadronic fluid. With the excitation energy higher then 140 MeV the boiled fluid evaporates the π – meson copiously. The spectra of the emitted π – mesons can be described by formula [5.11]:

$$d\eta = \frac{N_0 \frac{m}{T}}{K_2\left(\frac{m}{T}\right)} c^{-3}\gamma^5 u^2 \exp\left[-\frac{m\gamma}{T}\right] dV du. \tag{5.63}$$

In formula (5.63) $\gamma = \left(1 - \frac{u^2}{c^2}\right)^{-\frac{1}{2}}$, T is the temperature of the fluid and dV is the volume element. Function $K_2\left(\frac{m}{T}\right)$ is the modified Bessel function of the second kind.

5.4. Second Sound in Nuclear Interactions

The present paragraph is devoted to study of second sound propagation in nuclear matter. The velocity of heat propagation is calculated and the value $\upsilon_S = \upsilon_F / \sqrt{3}$ is obtained. Assuming the relaxation-time approximation, the diffusivity D and viscosity η are obtained. The second sound master equation is used to study the heating of two interacting nuclear slabs. The obtained solution of the equation describes two modes of heat transfer in nuclear matter: (i) ballistic propagation of the temperature pulse and (ii) heat diffusion. The temperature of the two interacting nuclear slabs is calculated.

Following paragraph 1.1 the heat transport equation in nuclear matter can be written as

$$\frac{1}{\upsilon_S^2}\frac{\partial^2 T}{\partial t^2} + \frac{1}{\tau \upsilon_S^2}\frac{\partial T}{\partial t} = \nabla^2 T. \tag{5.64}$$

Eq. (5.64) is the damped wave equation for the propagation of the second sound pulse with propagation velocity υ_s, which is the velocity of second sound in a nucleon Fermi gas. Assuming for the Fermi energy in the nucleus $E_F = 40$ MeV we obtain for the second sound velocity $\upsilon_s = 0.17\ c$, where c is the velocity of light. Considering that for a Fermi gas one can define the diffusivity D as:

$$D = \tau \upsilon_s, \tag{5.65}$$

where for nuclear matter the relaxation time $\tau \sim 10^{-23}$ s = 3 fm/c Eq. (5.64) can be written as

$$\frac{1}{\upsilon_S^2}\frac{\partial^2 T}{\partial t^2} + \frac{1}{D}\frac{\partial T}{\partial t} = \nabla^2 T \tag{5.66}$$

with $D = 8.7 \cdot 10^{-2}$ fm·c. For a Fermi gas the viscosity η is connected to diffusivity through formula

$$\eta = D\rho. \tag{5.67}$$

Assuming $D = 8.7 \cdot 10^{-2}$ fm·c and $\rho = 135.57$ MeV/(c^2fm^3) we obtain for the viscosity

$$\eta = 11.79 \text{ MeV/(fm}^2 c). \tag{5.68}$$

The obtained value of bulk viscosity η is in good agreement with presently adopted value of η for nuclear matter, $\eta = 3 - 15$ MeV c / fm.

Consider a cylindrical sample of nuclear matter with unit area which is heated at one end. The temperature at the other end of the sample is detected as a function of time. The purpose of this section is to discuss the solution of Eq. (5.64) which will be relate the signal at the temperature pulse detector (TP) $T(l, t)$ to the input pulse $T(0, t)$. The solution of Eq. (5.64) for a cylinder of infinite length is given by

$$T(x,t)=\frac{1}{2v_S}\int dx'T(x',0)\left[\begin{array}{l} e^{-t/2\tau}\frac{1}{t_0}\delta(t-t_0) \\ +e^{-t/2\tau}\frac{1}{2\tau}\left\{\begin{array}{l} I_0\left(\frac{\left(t^2-t_0^2\right)^{\frac{1}{2}}}{2\tau}\right) \\ +\frac{t}{\left(t^2-t_0^2\right)^{\frac{1}{2}}}I_1\left(\frac{\left(t^2-t_0^2\right)^{\frac{1}{2}}}{2\tau}\right)\end{array}\right\}\Theta(t-t_0)\end{array}\right] \quad (5.69)$$

where $t_0=(x-x')/v_S$ and I_0 and I_1 are modified Bessel functions. We concerned with solution to Eq. (5.66) when a nearly delta-function temperature pulse heats one end of the sample. Then at $t\sim 0$ the temperature distribution in the sample is

$$T(x,t)=\begin{cases}\Delta T_0 & \text{for } 0<x<v_S\Delta t,\\ 0 & \text{for } x>v_S\Delta t.\end{cases}$$

With this $t=0$ temperature profile, Eq. (5.64) yields

$$\begin{aligned}T(l,t)&=\tfrac{1}{2}\Delta T_0 e^{-t/2\tau}\Theta(t-t_0)\Theta(t_0+\Delta t-t)\\ &+\tfrac{1}{4}\Delta t\Delta T_0 e^{-t/2\tau}\left\{I_0(z)+\frac{t}{2\tau}\frac{1}{z}I_1(z)\right\}\Theta(t-t_0),\end{aligned} \quad (5.70)$$

where $z=\left(t^2-t_0^2\right)^{\frac{1}{2}}/2\tau$ and $t_0=l/v_S$. The first term in this solution corresponds to ballistic propagation of the second sound damped by $\exp(-t/2\tau)$ across the sample. The second term corresponds to the propagation of the energy scattered out of the ballistic pulse by diffusion. In the limit $\tau\to\infty$ the ballistic pulse alone arrives at the detector. In the limit $\tau\to 0$ the ballistic pulse is completely damped and the second term takes an asymptotic form, which is the solution to the conventional diffusion equation. In particular in this limit we have $z\to\infty$

$$I_0(z)\sim\frac{e^z}{(2\pi z)^{1/2}},\qquad I_1(z)\sim\frac{e^z}{(2\pi z)^{1/2}}, \quad (5.71)$$

so that the second term in Eq. (5.69) becomes

$$T(l,t) \sim 2\frac{\Delta t}{4\tau}\Delta T_0 \frac{e^{-\frac{t}{2\tau}} \exp\left[\frac{-\left(t^2 - t_0^2\right)^{\frac{1}{2}}}{2\tau}\right]}{\left[(\pi/\tau)\left(t^2 - t_0^2\right)^{\frac{1}{2}}\right]^{\frac{1}{2}}}. \tag{5.72}$$

Now, for $t >> t_0$, we can write $\left(t^2 - t_0^2\right)^{\frac{1}{2}} \sim t - \frac{1}{2}\left(t_0^2/t\right)$ and thus obtain

$$\lim_{\substack{t>>t_0 \\ \tau\to 0}} T(l,t) = \Delta T_0 \frac{\Delta t}{(4\pi\tau t)^{\frac{1}{2}}} \exp\left[-\frac{t_0^2}{4t\tau}\right]. \tag{5.73}$$

The solution to Eq. (5.66) when there are reflecting boundaries on the sample is the superposition of the temperature at l from the heated end and from image heat sources at $\pm 2nl$. This solution is

$$T(l,t) = \sum_{i=1}^{\infty}\left[\begin{array}{l}\Delta T_0 e^{-t/2\tau}\Theta(t-t_i)\Theta(t_i+\Delta t - t) \\ \qquad + \frac{\Delta t}{2\tau}\Delta T_0 e^{-t/2\tau}\left\{I_0(z_i) + \frac{t}{2\tau}\frac{1}{z_i}I_1(z_i)\right\}\Theta(t-t_i)\end{array}\right], \tag{5.74}$$

$$t_i = t_0, 3t_0, 5t_0, \ldots = l/v_S.$$

For short times only the $i = 0$ term in the sum is important. Each source begins to contribute at $x = l$ at the time (1) proportional to its distance (in the ballistic limit) or (2) proportional to the square of its distance (in the diffusion limit). After sufficiently long times one obtains always the diffusion limit. At the time $t_1 >> t_0$ all sources at a distance greater than l_1 given by

$$\left(\frac{l_1}{v_S\tau}\right)^2 \tau = \frac{(t_1)^2}{\tau} \tag{5.75}$$

do not contribute to the temperature at l at t_1. Each source which contributes to the temperature at l contributes the same amplitude

$$A_i = \Delta T_0 \frac{\Delta t}{(\pi\tau t_1)^{1/2}}, \tag{5.76}$$

so that the total heat at l is

$$T(l,t)=\sum_{i=1}^{N}\frac{\Delta T_0\Delta t}{(\pi\tau t_1)^{1/2}}, \tag{5.77}$$

where the number of contributing sources N is given by $(2N+1)l=l_1$ or $N\sim(\upsilon_S/l)\left[\frac{1}{2}(\tau t_1)\right]^{\frac{1}{2}}$ so that Eq. (5.77) becomes

$$T(l,t)\sim N\frac{\Delta T_0\Delta t}{(\pi\tau t_1)^{1/2}}\sim\left(\frac{\upsilon_S\Delta t}{l}\right)\Delta T_0, \tag{5.78}$$

which is proportional to the ration of the volume of the sample initially heated by the temperature source to the total volume.

In the diffusion limit $\tau\to 0$, $t>>t_0$ the asymptotic form of Eq. (5.74) is

$$T(l,t)=\sum_{i=1}^{N}\frac{\Delta T_0\Delta t}{(4\pi\tau t_1)^{1/2}}\exp\left[-\tfrac{t_i^2}{4\tau t}\right] \tag{5.79}$$

where the expansion of z_i in Eq.(5.74) is only valid for $t >> t_i$. Since the ith therm contributes to $T(l, t)$ only for $t\sim t_i^2/\tau$, it is valid to use the asymptotic expansion of each term. Eq. (5.79) is the solution of the diffusion equation (5.64) with $\tau = 0$, i.e.

$$\frac{\partial T}{\partial t}=\tau\upsilon_S^2\nabla^2 T. \tag{5.80}$$

This equation may also be solved in the form

$$T(x,t)=\frac{1}{l}\int_0^1 T(x,0)dx+\frac{2}{l}\sum_{n=1}^{\infty}\exp\left[-n^2\pi^2\tau t/t_0^2\right]\times\cos\frac{n\pi x}{l}\int_0^1 T(x',0)\cos\frac{n\pi x'}{l}dx', \tag{5.81}$$

where we have used the set of functions which naturally incorporate the boundary condition at the sample ends. For an initial temperature distribution such as (5.70) one finds [5.11]:

$$T(l,t)=\Delta T_0\frac{\upsilon_S\Delta t}{l}\left[1+2\sum_{n=1}^{\infty}(-1)^n\exp\left[-n^2\pi^2\tau t/t_0^2\right]\right]. \tag{5.82}$$

Note the prefactor is just the final temperature given by Eq. (5.78). In this form it is particularly convenient to see how to use measurements of $T(l, t)$ to learn the relaxation time τ. At long times, Eq. (5.82) is asymptotically:

$$T(l,t)\sim T_f\left[1-2\exp\left(-\pi\tau t/t_0^2\right)\right] \tag{5.83}$$

So that $T_f(l) - T(l, t)$ may be used to measure τ.

5.5. Production and Detection of Axion-Like Particles at VUV-FEL

New very light spin-zero particles which are very weakly coupled to ordinary matter are predicted in many models beyond the Standard Model. Such light particles arise if there is a global continuous symmetry in the theory that is spontaneously broken in the vacuum — a notable example being the axion [5.66], a pseudoscalar particle arising from the breaking of a U(1) Peccei-Quinn symmetry [5.67], introduced to explain the absence of *CP* violation in strong interactions. Such axion-like pseu-doscalars couple to two photons via

$$L_{\varphi\gamma\gamma} = -\frac{1}{4} g\varphi F_{\mu\nu}\widetilde{F}^{\mu\nu} = g\varphi\vec{E}\cdot\vec{B}. \tag{5.84}$$

where g is the coupling, φ is the field corresponding to the particle, $F_{\mu\nu}(\widetilde{F}^{\mu\nu})$ is the (dual) electromagnetic field strength tensor, and $\vec{E}$ and $\vec{B}$ are the electric and magnetic fields, respectively. In the case of a scalar particle coupling to two photons, the interaction reads

$$L_{\varphi\gamma\gamma} = -\frac{1}{4} g\varphi F_{\mu\nu}F^{\mu\nu} = g\varphi\left(\vec{E}^2 - \vec{B}^2\right) \tag{5.85}$$

Both effective interactions give rise to similar observable effects. In particular, in the presence of an external magnetic field, a photon of frequency ω may oscillate into a light spin-zero particle of small mass $m_\varphi<\omega$, and vice versa. The notable difference between a pseudoscalar and a scalar is that it is the component of the photon polarization parallel to the magnetic field that interacts in the former case, whereas it is the perpendicular component in the latter case.

The exploitation of this mechanism is the basic idea behind photon regeneration (sometimes called "light shining through walls") experiments [5.68, 5.69]. Namely, if a beam of photons is shone across a magnetic field, a fraction of these photons will turn into (pseudo-) scalars. This (pseudo-)scalar beam can then propagate freely through a wall or another obstruction without being absorbed, and finally another magnetic-field located on the other side of the wall can transform some of these (pseudo-)scalars into photons — apparently regenerating these photons out of nothing. A pilot experiment of this type was carried out in Brookhaven using two prototype magnets for the Colliding Beam Accelerator [5.70]. From the non-observation of photon regeneration, the Brookhaven-Fermilab-Rochester-Trieste (BERT) collaboration excluded values of the coupling $g < 6.7\times10^{-7}$ GeV^{-1} for $m_\varphi \leq 10^{-3}$ eV [5.71].

Recently, the PVLAS collaboration lias reported an anomalous signal in measurements of the rotation of the polarization of photons in a magnetic field [5.72]. A possible explanation of such an apparent vacuum magnetic dichroism is through the production of a light pseudoscalar or scalar, coupled to photons through Eq. (5.84) or Eq. (5.85), respectively. Accordingly, photons polarized parallel (pseudoscalar) or perpendicular (scalar) to the magnetic field disappear, leading to a rotation of the polarization plane [5.73]. The region quoted in Ref. [5.72] that might explain the observed signal is

$$1.7\times10^{-6}\ \text{GeV}^{-1} < g < 5.0\times10^{-6}\ \text{GeV}^{-1}, \tag{5.86}$$

$$1.0\times10^{-3}\ \text{eV} < m_{\varphi} < 1.5\times10^{-3}\ \text{eV}, \tag{5.87}$$

obtained from a combination of previous limits on g vs. m_{φ} from a similar, but less sensitive polarization experiment performed by the BFRT collaboration [5.71] and the g vs. m_{φ} curve corresponding to the PVLAS signal.

A particle with these properties presents a theoretical challenge. It is hardly compatible with a genuine QCD axion. Moreover, it must have very peculiar properties in order to evade the strong constraints on g from stellar energy loss considerations [5.74] and from its non-observation in helioscopes such as the CERN Axion Solar Telescope [5.75, 5.76]. Its production in stars may be hindered, for example, if the $\varphi\gamma\gamma$ vertex is suppressed at keV energies due to low scale compositeness of φ, or if, in stellar interiors, φ acquires an effective mass larger than the typical photon energy, ~ keV, or if the particles are trapped within stars [5.77 – 5.79].

Clearly, an independent and decisive experimental test of the pseudoscalar interpretation of the PVLAS observation, without reference to axion production in stars (see [5.80, 5.81]), is urgently needed. In Ref. [5.82], one of us (AR) was involved in the consideration of the possibility of exploiting powerful high-energy free-electron lasers (FEL) in a photon regeneration experiment to probe the region where the PVLAS signal could be explained in terms of the production of a light spin-zero particle. In particular, it was emphasized that the free-electron laser VUV-FEL [5.83] at DESY, which is designed to provide tunable radiation from the vacuum-ultraviolet (VUV: 10 eV) to soft X-rays (200 eV), will offer a unique and timely opportunity to probe the PVLAS result.

Table 5.4. Achieved (2005) and expected (2007) VUV-FEL parameters.

	2005	2007
Bunch separation [ns]	1000	1000
Bunches per train #	30	800
Repetition rate [1/s]	5	10
Photon wavelength [nm]	32	32
Photon energy [eV]	38.7	38.7
Energy per pulse [μJ]	10	50
Photons per pulse #	1.6 x 10^{12}	8.1 x 10^{12}
Average flux [1/s]	2.4 x 10^{14}	6.5 x 10^{16}

Notably, the high photon energies available at the VUV-FEL increase substantially the expected photon regeneration rate in the mass range implied by the PVLAS anomaly, in comparison to the one expected at visible (~ 1 eV) lasers.

The proposed experiment is based on the assumption that the VUV-FEL can deliver photons with an energy ω = 38.7 eV and an average photon flux $\dot{N}_0 = 6.5\times10^{16}\,\mathrm{s}^{-1}$ (cf. Table 5.4). For the proposed photon regeneration experiment at the VUV-FEL, we study a linear arrangement of 12 normal conducting dipole magnets which are freely available at DESY. Each of these magnets has a magnetic field of 2.24 T and an integrated magnetic length of 1.029 m. The default arrangement consists of six plus six magnets, the beam absorber being placed between the first and second six. This arrangement corresponds to a magnetic field region of size $BL = 2Bl = 27.66$ Tm. The proposed configuration is too large to fit into the VUV-FEL experimental hall. It has to be built on the ground before the entrance. Correspondingly, the FEL beam line has to be extended to the proposed experiment.

The photons leave the VUV-FEL with horizontal linear polarization. In order to have a maximal coupling with a possible pseudoscalar/scalar, the magnetic field $\vec{B}$ of the magnets before the absorber should lie in the horizontal/perpendicular direction. We therefore foresee to exploit both possibilities of the magnetic field direction.

For the proposed experiment, the expected flux of regenerated photons is

$$\dot{N}_f \approx 1\times10^{-4}\,\mathrm{s}^{-1}F^2(ql)\left(\frac{\dot{N}_0}{6.5\times10^{16}\,\mathrm{s}^{-1}}\right)\times\left(\frac{g}{10^{-6}\,\mathrm{GeV}^{-1}}\right)^4\left(\frac{B}{2.24\,\mathrm{T}}\right)^4\left(\frac{l}{6\,\mathrm{m}}\right)^4, \quad (5.88)$$

where $q = m_\varphi^2/(2\omega)$ $(<< m_\varphi)$ is the momentum transfer magnet and

$$F(ql) = \left[\frac{\sin\left(\frac{1}{2}ql\right)}{\frac{1}{2}ql}\right]^2 \quad (5.89)$$

is a form factor which reduces to unity for small ql, corresponding to large ω; or small m_φ,

$$m_\varphi << \sqrt{\frac{2\pi\omega}{l}} = 3\times10^{-3}\,\mathrm{eV}\sqrt{\left(\frac{\omega}{38.7\,\mathrm{eV}}\right)\left(\frac{6\,\mathrm{m}}{l}\right)}. \quad (5.90)$$

For smaller ω or larger m_φ, incoherence effects set in between the (pseudo-)scalar and the photon, the form factor getting much smaller than unity, severely reducing the regenerated photon flux (5.88). Therefore, a photon regeneration experiment exploiting the VUV-FEL beam has a unique advantage when compared to one using an ordinary laser operating near the visible (ω ~ 1 eV): the sensitivity of the former extends to much larger masses. In particular, the mass range (5.87) implied by PVLAS is entirely covered, for ω = 38.7 eV photons (cf. Eq. (5.90)).

In case a signal is found, one may use the possibility to tune the photon energy for a determination of the mass of the particle. This is done by lowering the energy of the FEL photons and observing the on-set of the incoherence expected around $\omega \approx m_{\varphi}^{2} l / 2\pi$. A determination of the form factor F as a function of ω; will allow an extraction of m_{φ}.

The flux prediction (5.88) uses the benchmark values for the VUV-FEL flux and for the proposed magnetic field arrangement. For g and m_{φ} in the parameter region preferred by PVLAS, Eqs. (5.86) and (5.87), this results in a rate of regenerated photons ranging from about 1 mHz up to 1 Hz.

The very low predicted rates require therefore a detector system with

- a large single photon efficiency at ~ 40 eV,
- a short response time,
- and a low noise rate.

Three detector options are being considered: electron multipliers, multi-channel plates, and avalanche photo diodes. One manufacturer quotes an efficiency of about 7% for electron multipliers. For the two other options, extrapolations point to a value of around 10%. All three detectors show a response time in the 10 ns range. This short response time allows a reduction of the noise rate by timing, exploiting the time structure of the photon beam. These detector performances, in particular the efficiencies and the response time, have to be studied at a beamline of the VUV-FEL as soon as possible. At the same time, the general background rates in the VUV-FEL environment have to be studied as well.

In the case of a non-observation of photon regeneration, for an assumed running time of 12 x 12 h with an average photon flux of $\dot{N}_0 = 6.5 \times 10^{16}\ \mathrm{s}^{-1}$ at ω = 38.7 eV, a 7% single photon efficiency and zero background, the proposed experiment can establish a 95% confidence limit of $g < 8.8 \times 10^{-7}\ \mathrm{GeV}^{-1}$, for $m_{\varphi} \leq 3 \times 10^{-3}\ \mathrm{eV}$. In particular, the experiment is expected to be able to firmly exclude the particle interpretation of the PVLAS anomaly and to improve the current laboratory bound on g in the $m_{\varphi} \geq 10^{-3}\ \mathrm{eV}$ range.

The proposed experiment offers a window of opportunity for a firm establishment or exclusion of the particle interpretation of the PVLAS anomaly in the near future. It takes essential advantage of unique properties of the VUV-FEL beam. The available VUV-FEL photon energies are just in the range where the photon regeneration rate is most sensitive to the hypothetical particle's mass. Moreover, the well-defined beam of the VUV-FEL will not produce beam-related backgrounds.

The experiment should be done soon, before other experiments [5.84, 5.85] can compete. A first step towards this goal is the study of possible detectors and their background rates. Finally, the proposed experiment could serve also as a test facility for an ambitious large scale photon regeneration experiment [5.86].

REFERENCES

[5.1]. Kaplan, A. E. *Phys. Rev. Lett.* 1994, *73*, 1243.

[5.2]. Paul, P. M.; et al. *Science* 2001, *292*, 1689. Hertz E.; et al. *Phys. Rev.* 2001, *A 64*, 051801.
[5.3]. Hartemann, F. V. *Phys. Plasmas* 1998, *5*, 2037.
[5.4]. Yu, W.; et al. *X-ray Lasers—1996,* IOP Conf. Proc. No. 151; Institute of Physics and Physical Society: London, 1996; 460.
[5.5]. Sarachik, E. S.; Shappert, G. T. *Phys. Rev.* 1970, *D1*, 2738.
[5.6]. Kaplan, A. E. *Phys. Rev. Lett.* 1986, *56*, 456.
[5.7]. Shen, B. F.; Meyer-ter-Vehn, J. *Phys. Plasmas* 2001, *8*, 1003.
[5.8]. Landau, L.; Lifshitz, E. *Classical Field Theory*; Pergamon: New York, NY, 1975; Jackson, J. D. *Classical Electrodynamics*; Wiley: New York, NY, 1975.
[5.9]. Kimel, I.; Elias, L. R. *Phys. Rev. Lett.* 1995, *75*, 4210; *Nucl. Instnim. Methods Phys. Res., Sect.* 1996, *A 375*, 565.
[5.10.]Kaplan, A. F.; Shkolnikov, P. *Phys. Rev. Lett.*, 2002, *88*, 074801-1.
[5.11].Kozlowski, M.; Marciak – Kozlowska, J. *From Quarks to Bulk Matter*; Hadronic Press: Palm Harbor, FL, 2001.
[5.12].Calogero, F. *Phys. Letters,* 1997, *A228*, 335.
[5.13].Kozlowski, M.; Marciak – Kozlowska, J. *Nuovo Cimento* 2001, *116B*, 821.
[5.14].Anderson, J. D.; et al. *Phys. Rev. Lett.* 1998, *81*, 2858.
[5.15].Marciak – Kozlowska, J.; Kozlowski, M. *Foundations of Physics Letters* 1996, *9,* 235.
[5.16].Kozlowski, M.; Marciak – Kozlowska, J. *Foundations of Physics Letters,* 1997, *10*, 295.
[5.17].Kozlowski, M.; Marciak – Kozlowska, J. (2003). Aging of the Universe and the fine structure constant. arXiv/astro-ph/0307 168.
[5.18].Planck, M. *The theory of heat radiation;* Dover Publications: New York, NY, 1959; 173.
[5.19].Yukawa, H. *Proc. Phys.-Math. Soc. Japan* 1935, *17*, 48.
[5.20].Schwinger, J. *Phys. Rev.* 1951, *82*, 664.
[5.21].Artur, J.; et al. *SLAC-R-0521* 1998.
[5.22].Brinkmann, R.; et al. *Proc. IEEE Particle Acceleration Conference JUPAP* 1995, 674.
[5.23].Materlik, G.; Tschentscher, T. *TESLA: The superconducting electron positron linear collider with an integrated X-ray laser laboratory, Technical design report, Pt. 5: The X-ruyfree electron laser,* DESY-01-011, DESY-2001-011, DESY-01-OllE, DESY-2001-011E, DESY-TESLA-2001-23, DESY-TESLA-FEL-2001-05, ECFA-2001-209.
[5.24].Tajima, T. "Fundamental Physics with an X-Ray R-ee Electron Laser", subm. to *Comments Plasma Phys. Contr. Fusion;* Tajima T.; Mourou, G. (2001). Zettawatt – Exqwatt laser and their application in ultrastrong – field physics. arXiv/physics/0111091.
[5.25].Melissinos, A. C. In *Quantum Aspects of Beam Physics;* Chen, P.; World Scientific: Singapore, 1998; 564.
[5.26].Chen, P.; Pellegrini, C. In *Quantum Aspects of Beam Physics;* Chen, P.; World Scientific: Singapore, 1998; 571.
[5.27].Ringwald, A. *Phys. Lett.* 2001, *B 510*, 107.
[5.28].Alkofer, R.; et al. *Phys. Rev. Lett.* 2001, *87*, 193902.
[5.29].Popov, V. S. *Pisma Zh. Eksp. Tear. Fiz.* 2001, *74*, 151.
[5.30].Chen, P.; Tajima, T. *Phys. Rev. Lett.* 1999, *83*, 265.
[5.31].Rosu, H. C. *Phys. World* 1999, *10*, 21;
[5.32].Melissinos, A. C. *Nucl. Phys. Proc. Suppl.* 1999, *72*, 195.
[5.33].Ringwald, A. (2007). Particle interpretations of the PVLAS Data. arXiv:0704.3195.
[5.34].Madey, J. M. *J. Appl. Phys.* 1971, *42*, 1906.

[5.35]. Kondratenko, A. M.; Saldin, E. L. *Part. Accel.* 1980, *10*, 207; Bonifacio, R.; Pellegrini, C.; Narducci, L. M. *Opt. Common.* 1984, *50*, 373.
[5.36]. Andruszkow, J.; et al. *Phys. Rev. Lett.* 2000, *85*, 3825.
[5.37]. F. Sauter, *Z. Phys.* 1931, *09,* 742; Heisenberg, W.; Euler, H. *Z. Phys.* 1936, *98*, 714; Schwinger, J. *Phys. Rev.* 1951, *82*, 664.
[5.38]. Hawking, S. W. *Nature* 1974, *248,* 30; *Commun. Math. Phys.* 1975, *43*, 199; Damour, T.; Ruffini, R. *Phys. Rev.* 1976, *D 14*, 332; Gibbons G. W.; Perry, M. J. *Proc. Roy. Soc. Land.* 1978, *A 358*, 467; Gavrilov, S. P.; Gitman, D. M. *Phys. Rev.* 1996, *D 53*, 7162; Parikh M. K.; Wilczek, F. *Phys. Rev. Lett.* 2000, *85*, 5042.
[5.39]. Casher, A.; Neuberger, H.; Nussinov, S. *Phys. Rev.* 1979, *D 20*, 179; Andersson, B.; Gustafson, G.; Ingelman, G.; Sjdstrand, T. *Phys. Rept.* 1983, *97*, 31; Biro, T. S.; Nielsen, H. B.; Knoll, J. *Nucl. Phys.* 1984, *B 245*, 449.
[5.40]. Parker, L. *Phys. Rev.* 1969, *183*, 1057; Birrell, N. D.; Davies, P. C. *Quantum Fields in Curved Space.* Cambridge University Press, 1982.
[5.41]. Greiner, W.; Muller, B.; Rafelski, J. *Quantum Electrodynamics of Strong Fields.* Springer-Verlag, Berlin, 1985; Grib, A. A.; Mamaev, S. G.; Mostepanenko, V. M. *Vacuum Quantum Effects in Strong Fields.* Atomizdat, Moscow, 1988; Fradkin, E. S.; Gitman, D. M.; Shvartsman, Sh. M. *Quantum Electrodynamics with Unstable Vacuum.* Springer-Verlag: Berlin, 1991.
[5.42]. Bunkin, F. V.; Tugov, I. I. *Dokl. Akad. Nauk Ser. Fiz.* 1969, *187*, 541.
[5.43]. Brezin, E.; Itzykson, C. *Phys. Rev.* 1970, *D 2*, 1191.
[5.44]. Popov, V. S. *Pisma Zh. Eksp. Tear. Fiz.* 1971, 13, 261; Popov, V. S. *Zh. Eksp. Tear. Fiz.* 1971, *61*, 1334.
[5.45]. Troup, G. J.; Perlman, H. S. *Phys. Rev.* 1972, *D 6*, 2299; Popov, V. S. *Zh. Eksp. Tear. Fiz.* 1972, *62*, 1248; Popov, V. S.; Marinov, M. S. *Yad. Fiz.* 1972, *16*, 809; Narozhnyi, N. B.; Nikishov, A. I. *Zh. Eksp. Tear. Fiz.* 1973, *65*, 862.
[5.46]. Popov, V. S. *Pisma Zh. Eksp. Tear. Fiz.* 1973, *18*, 435.
[5.47]. Mostepanenko, V. M.; Frolov, V. M. *Yad. Fiz.* 1974, *19*, 885.
[5.48]. Popov, V. S. *Yad. Fiz.* 1974, *19*, 1140.
[5.49]. Marinov, M. S.; Popov, V. S. *Fortsch. Phys.* 1977, *25*, 373.
[5.50]. Katz, J. I. *Astrophys. J. Supp.* 2000, *127*, 371.
[5.51]. Dunne, G.; Hall, T. *Phys. Rev.* 1998, *D 58*, 105022.
[5.52]. Fried, H. M.; et al. *Phys. Rev.* 2001, *D 63*, 125001.
[5.53]. Paul, P. *Science* 2001, *292*, 1689.
[5.54]. Hentchel M.; et al. *Nature* 2001, *414,* 509.
[5.55]. Audebert P.; et al. *TESLA XFEL: First stage of the X-ray laser laboratory - Technical design report.* Brinkmann, R.; et al.; Preprint DESY 2002-167.
[5.56]. The LCLS Design Study Group, LCLS Design Study Report, SLAC reports SLAC-R521, Stanford, 1998.
[5.57] Saldin, E. L.; Schneidmiller E. A.; Yurkov, M. V. *DESY 04-013.* 2004.
[5.58]. Saldin, E. L.; Schneidmiller, E. A.; Yurkov, M. V. (2004). A new technique to generate 100 GW – level attosecond X – ray pulses from the X – ray SASE FELs. arXiv:0403067.
[5.59]. Saldin, E. L.; Schneidmiller E. A.; Yurkov, M. V. *The Physics of Free Electron Lasers*; Springer-Verlag: Berlin, 1999.
[5.60]. Nelson, E. *Phys. Rev.* 1966, *150*, 1079.
[5.61]. Catteneo, C., *Atti. Sem. Mat. Fis. Univ. Modena* 1948, *3*, 3; *Compte. Rendus* 1958, *247*, 431.
[5.62]. Luikov, A. V. *Analytical Heat Diffusion Theory*; Academic Press: New York, NY, 1968.

[5.63]. Joseph, D. D.; Preziosi, L. *Rev. Mod. Phys.* 1989, *61*, 41.
[5.64]. Jou, D.; et al. *Extended Irreversible Thermodynamics*; Springer-Verlag: Berlin, 1993.
[5.65]. Perkins, D. H. *Introduction to High Energy Physics*; Addison-Wesley: Menlo Park, Ca, 1987.
[5.66]. Weinberg, S. *Phys. Rev. Lett.* 1978, *40*, 223; Wilczek, F. *Phys. Rev. Lett.* 1978, *40*, 279.
[5.67]. Peccei, R. D.; Quinn, H. R. *Phys. Rev. Lett.* 1977, *38*, 1440; Peccei, R. D.; Quinn, H. R. *Phys. Rev.* 1977, *D 16*, 1791.
[5.68]. Sikivie, P. *Phys. Rev. Lett.* 1983, *51*, 1415; Anselm, A. A. *Yad. Fiz.* 1985, *42*, 1480; Gasperini, M. *Phys. Rev. Lett.* 1987, *59*, 396.
[5.69]. Van Bibber, K.; et al. *Phys. Rev. Lett.* 1987, *59*, 759.
[5.70]. Ruoso, G.; et al. *Z. Phys.* 1992, *C 56*, 505.
[5.71]. Cameron, R.; et al. *Phys. Rev.* 1993, *D 47*, 3707.
[5.72]. Zavattini, E.; et at. *Phys. Rev. Lett.* 2006, *96*, 110406.
[5.73]. Maiani, L.; Petronzio, R.; Zavattini, E. *Phys. Lett.* 1986, *B 175*, 359; Raffelt, G.; Stodolsky, L. *Phys. Rev.* 1988, *D 37*, 1237.
[5.74]. Raffelt, G. G. *Phys. Rev.* 1986, *D 33*, 897.Raffelt G. G.; Dearborn, D. S. *Phys. Rev.* 1987, *D 36*, 2211; Raffelt, G. G. *Stars As Laboratories For Fundamental Physics: The Astrophysics of Neutrinos, Axions and other Weakly Interacting Particles*; University of Chicago Press: Chicago, IL, 1996; Raifelt, G. G. *Ann. Rev. Nucl. Part. Sci.* 1999, *49*, 163.
[5.75]. Zioutas, K.; et al. *Phys. Rev. Lett.* 2005, *94*, 121301.
[5.76] Raffelt, G. G. In *Proceedings of the Eleventh International Workshop on "Neutrino Telescopes"*; Milla Baldo Ceolin; Istituto Veneto di Scienze, Lettere ed Arti: Campo Santo Stefano, 2005; 419-431.
[5.77]. Masso, E.; Redondo, J. *JCAP* 2005, *0509*, 015.
[5.78]. Jain, P.; Mandal, S. (2005). Evading the astrophysical limits on light pseudoscalars. arXiv:astro-ph/0512155.
[5.79]. Jaeckel, E.; et al. TITLE OF PAPER. arXiv:hep-ph/0605313.
[5.80]. Dupays, A.; et al. *Phys. Rev. Lett.* 2005, *95*, 21130205.
[5.81]. Kleban, M.; Rabadan, R. (2005). Collider Bounds on Pseudoscalars Coupling to Gauge Bosons. arXiv:hep-ph/0510183.
[5.82]. Rabadan, R.; Ringwald, A.; Sigurdson, K. *Phys. Rev. Lett.* 2006, *96*, 110407.
[5.83]. Free-electron laser at TESLA Test Facility,http://www-hasylab.desy.de/facility/fel/main.htm;
[5.84]. Cantatore G.; et al. (2006). PVLAS: recent results. DESY seminar. http://www.desy.de/f/seminar/Cantatore.pdf
[5.85]. Battesti, R. (2005). The Birefringence Magnetique du Vide project: axion search with a pulsed magnet - Status of the experiment. http://cast.mppmu.mpg.de/axion-training-2005/axion-trainini
[5.86] Ringwald, A. *Phys. Lett.* 2003, *B 569*, 51.

Chapter 6

BEYOND ORTHODOX QUANTUM THEORY

The usual interpretation of the quantum theory requires us to give up the concepts of causality, continuity, and the objective reality of individual micro-objects, in connection with the quantum-mechanical domain. Instead it leads to a point of view in which physics is said to be inherently and unavoidably restricted, in this domain and below, to the manipulation of mathematical symbols according to suitable techniques that permit, in general, the calculation only of the *probable* behaviour of the phenomena that can be observed in the macroscopic domain. These far-reaching changes in the conceptual structure of physics have been based on the assumption that certain features of the current formulation of the quantum theory, viz. the indeterminacy principle and the appearance of a characteristic set of opposing "complementary" pairs of modes of behaviour (e.g. wave-like and particle-like), are absolute and final features of the laws of nature, which will continue to apply, uncontradicted and without approximation, in every domain that will ever be the subject of physical investigation.

David Bohm [6.1] has demonstrated that it is not necessary to make this assumption, and that indeed such an assumption constitutes a dogmatic restriction on the possible forms of future theories. In the present chapter, we shall present some specific theory which allows us to interpret the quantum mechanics in a new way. This theory permits the representation of quantum-mechanical effects as arising out of an objectively real sub-stratum of continuous motion, existing at a lower level, and satisfying new laws which are such as to lead to those of the current quantum theory as approximations that are good only in what we shall call the quantum-mechanical level. The new theory serves two principal purposes. First of all, it help to put into a more specific form the criticisms of the usual interpretation of the quantum mechanics (Bohr, Jordan, Born interpretation) Secondly, and perhaps even more important, the new theory (D. Bohm interpretation) may serve as useful starting-points in investigations aimed at the understanding of new domains of phenomena that are not yet very well understood. In connection with the second point mentioned above, let us recall that there now exists a crisis in physics, arising from the inadequacy of current theories in the treatment of phenomena involving very high energies and very short distances (of the order of 10^{-15}m or less). Of course, the proponents of the usual interpretation of the quantum theory are, on the whole, quite aware of this crisis. Consider, for example, nineteenth-century physicists could equally well have claimed that the unbroken success of the deterministic laws of classical mechanics in three centuries of applications was a very strong indication that progress into new domains would be very likely to lead to laws that were, if anything, even more deterministic than those that already existed. (In fact, many physicists of the time did hope that the laws of classical statistical mechanics would eventually be deduced completely and perfectly from a deterministic basis.) Thus, it would seem that historical experience should teach us not to make simple extrapolations of previous trends, when we come to the question of what is the degree to which the laws of new domains will show a statistical or a precisely determinate character. Rather it seems clear that one should not decide this question *a priori,* but, instead, one should be ready to try various kinds of laws and to see which kind permits the greatest progress in the understanding of the new domains.

We note, first of all, that if one adopts the hypothesis of a sub-quantum mechanical level containing hidden variables, then, as pointed by D. Bohm we are led to regard the statistical character of the current quantum theory as originating in random fluctuations of new kinds of entities, existing in the lower level. If we consider only those entities which can be defined at the quantum-mechanical level alone, these will be subjected to a genuine indeterminacy in their motions, because determining factors that are important (i.e. the hidden variables) simply cannot be defined in this level. Hence, as in the usual interpretation of the quantum theory, we regard the indeterminacy implied by Heisenberg's principle as an objective necessity and not just as a consequence of a simple lack of knowledge on our part concerning some hypothetical "true" states of the quantum-mechanical variables. Thus, it is not the existence of indetermination and the need for a statistical theory that distinguishes our point of view from the usual one. For these features are common to both points of view. The key difference is that we regard this particular kind of indeterminacy and the need for this particular kind of statistical treatment as something that exists only within the context of the quantum-mechanical level, so that by broadening the context we may diminish the indeterminacy below the limits set by Heisenberg's principle.

To go beyond the limits set by Heisenberg's principle, it will be necessary to use new kinds of physical processes that depend significantly on the details of what is happening at the sub-quantum mechanical level. As we shall see later, there is some reason to believe that such processes could perhaps be found in the domain of very high energies very short distances and very short times. It is clear, however, that in any process which can be treated to an adequate degree of approximation by the laws of the current quantum theory, the entities existing in the lower level cannot be playing any very significant role. Very little information about these entities could then be obtained by observing the results of this kind of process. In such an observation, Heisenberg's principle would, therefore, apply to a very high degree of approximation as a correct limitation on how well the state of an individual physical system could be determined, while evidently, if we observed the system with the aid of physical processes sensitive to the precise states of the hidden variables, this limitation would cease to be applicable.

To illustrate in more detail what the indeterminacy principle would mean in terms of a sub-quantum mechanical level, it will be helpful to return here to the analogy of Brownian movement

As we know, the motion of a smoke particle is subject to random fluctuations, originating in collisions with the atoms which exist at a lower level. As a result of these collisions, its motions cannot be completely determined by any variables (e.g. the position and velocity of the particle) existing at the level of the Brownian motion itself. Indeed, the lack of determination is not only qualitatively analogous to that obtained in the quantum theory, but, as has been shown by Furth [6.2] the analogy even extends to the quantitative form of the indeterminacy relations. Thus if we observe a moving smoke particle throughout some short interval of time, Δt we will find random fluctuations of magnitude Δx in the mean position, and of magnitude, ΔP in its mean momentum, which satisfy the relationship!

$$\Delta P \Delta x \approx C.$$

Here C is a constant, which depends on the temperature of the gas, as well as on other properties such as its viscosity. The form of this relationship is just the same as that of Heisenberg, except that the Planck's constant, h, has been replaced by the constant, C, which depends on the state of the gas.

There is, however, an important respect in which the analogy between the Brownian motion and the quantum theory is not complete. This difference arises essentially in the fact that C is not a universal constant whereas h is. As a result, in principle at least, one is able by changing conditions suitably to make C arbitrarily small (e.g. by lowering the temperature) and thus reduce the indeterminacy below any desired value. On the other hand, the constant, h does not depend on conditions in any known way, so that Heisenberg's relations imply, as far as we have been able to tell, an indeterminacy that is universal, at least within the quantum-mechanical domain. This means that while we can by a suitable choice of conditions construct apparatus (e.g. a microscope) which is not significantly affected by the kind of Brownian motion that we wish to observe, we cannot obtain a similar result in connection with the quantum-mechanical indeterminacy. To improve the analogy, we should therefore have to suppose that in the quantum domain we are effectively restricted to using apparatus that is itself undergoing the Brownian motion to an extent that is comparable with that undergone by the micro-systems that we are trying to observe. If we recall, however, that in our point of view, *all* matter in all its known manifestations is continually undergoing fluctuations originating in the sub-quantum mechanical level, we can see that the above extension of the analogy is justifiable. Considering that these fluctuations are present everywhere with essentially the same characteristics, we conclude that the universal and uniform character of the limitations implied by Heisenberg's principle in the quantum domain would not be an unexpected consequence of our assumptions.

To overcome these limitations we should have to take advantage of properties of matter that depended significantly on the sub-quantum mechanical level. One way to do this would be to make our observations with the aid of processes that were *very fast compared* with the sub-quantum mechanical fluctuations, so that the whole measurement would be over before these fluctuations could have changed things by very much. Such rapid processes are most likely to be obtained in the high energy domain since, from the Einstein relation, $E= h\nu$, a high energy, E, implies a process of high frequency, ν.

Finally, the analogy of Brownian motion also serves to bring out two different limiting modes in which the indeterminacy originating in random sub-quantum mechanical fluctuations may manifest itself. For let us consider, not the Brownian motion of smoke particles, but rather that of very fine droplets of mist. It is evident that there is a certain indeterminacy in the motion of these droplets that could be removed only by going to a broader context, including the air molecules with which these droplets are continually being struck. It remains true, however, that in their irregular Brownian motions the droplets retain their characteristic mode of existence as very small bodies of water. On the other hand, as we approach the critical temperature and pressure of the gas a new behaviour appears; for the fine droplets begin to become unstable. The substance then enters a phase in which the droplets are always forming and dissolving and, as a result, the substance becomes opalescent.

Here we have a new kind of fluctuation, which leads to an indeterminacy in the very mode of existence of the substance (i.e. between existence in the form of droplets and existence in the form of a homogeneous gas).

Similarly, it is possible that the very mode of existence of the electron will eventually be found to be indeterminate, when we have understood the detailed character of quantum fluctuations. Indeed, the fact that the electron shows a characteristic wave particle duality in its behaviour would, suggest that perhaps this second kind of indeterminacy will turn out to be the relevant one; for if such an indeterminacy exists, it would lead to a concept of the electron as an entity that was continually fluctuating from wave-like to particlelike character, and thus capable of demonstrating both modes of behaviour, each of which would, however, be emphasized differently in the different kinds of environment supplied, for example, by different arrangements of laboratory apparatus.

Of course, we have no way at present to decide which of these interpretations of the indeterminacy principle is the correct one. Such a decision will be possible only when we shall have found an adequate theory that goes below the level of the quantum theory. Meanwhile, however, it is important to keep both possibilities in mind. In the subsequent work, we shall therefore discuss examples of both kinds of theories. It is significant to note that the first steps towards an alternative interpretation of the quantum theory were taken about hundred years ago by de Broglie [6.3, 6.4] and by Madelung [6.5] at more or less the same time as the usual interpretation was being brought into its current definitive form. Neither of these steps was, however, carried far enough to demonstrate the possibility of a consistent treatment of all the essential aspects of the quantum theory. Indeed, the interpretation of de Broglie was subjected to severe criticisms by some of the proponents of the usual interpretation. Partly as a result of these criticisms and partly as a result of additional criticisms which he made himself, de Broglie gave up his proposals for a long time.

After these efforts had died out, it was not until about 1950 [6.1] that a systematic tendency to question the usual interpretation of the quantum theory began to develop on an appreciable scale. Among the most thoroughgoing of the earlier critical efforts in this direction were those of Blokhinzhev and Terletzky [6.6]. These physicists made it clear that it is not necessary to adopt the interpretation of Bohr and Heisenberg, and showed that instead, one may consistently regard the current quantum theory as an essentially statistical treatment, which would eventually be supplemented by a more detailed theory ermitting a more nearly complete treatment of the behaviour of the individual system. They did not, however, actually propose any specific theories or models for the treatment of the individual systems. Then in 1951, D. Bohm began to seek such a model; and indeed shortly thereafter he found a simple causal explanation of the quantum mechanics which, as he later learned, had already been proposed by de Broglie in 1927. Meanwhile, however, the theory had been carried far enough so that the fundamental objections that had been raised against the suggestions of de Broglie had been answered. This was done mainly with the aid of a theory of measurements which showed that the new interpretation was consistent with all the essential characteristics of the quantum theory. Partly as a result of this work, and partly as a result of additional suggestions made by Vigier and de Broglie [6.7] then returned to his original proposals, since he now felt that the decisive objections against them had been answered.

At this stage, D. Bohm principal purpose had not been to propose a definitive new theory, but was rather mainly to show, with the aid of a concrete example, that alternative interpretations of the quantum theory were in fact possible.

We shall now discuss briefly the current crisis in microscopic physics, in order to lay a foundation for an explanation of some of the advantages of the point of view towards the quantum theory that we have developed in this chapter.

Let us first discuss the theoretical aspects of this crisis. When one applies the existing quantum theory to the electrodynamics of "elementary" particles (such as electrons, protons, etc.), internal inconsistencies seem to arise in the theory. These inconsistencies are connected with the prediction of infinite values for various physical properties, such as the mass and the charge of the electron. AH these infinities arise from the extrapolation of the current theory to distances that are unlimitedly small. Among the things that make such an extrapolation necessary, one of the most important is the assumption, which seems to be an intrinsic part of current theories that "elementary" particles, such as electrons, are mathematical points in the sense that they occupy no space at all. On the other hand, in spite of many years of active searching on the part of theoretical physicists throughout the whole world, no way has yet been found to incorporate consistently into the current quantum theory the assumption that the electron occupies a finite region of space. While it has been suggested that perhaps the infinities come from an inadequate technique of solving the equations (i.e. perturbation theory), persistent efforts to improve this technique have not yet produced any favourable results, and indeed those results that have been obtained favour the conclusion that basically it is not the mathematical technique that is at fault, but rather the theory itself is not logically consistent.

Within the framework of the Bohm's theory, it is still possible to calculate many results, namely, those which do not depend critically on the assumed size of the particle. A few years ago, important new successes in this direction were obtained by Tomanaga [6.8], Schwinger [6.9] and Feyman [6.10], with the prediction of certain very fine details of the spectrum of hydrogen gas, as well as with experiments that measured the magnetic moment of the electron. Impressive as these results are, considered as examples of extraordinarily complex calculations that led to correct experimental predictions, they throw little light, however, on the problem of the infinities that is one of the most important manifestations of the current crisis in physics. For a closer examination of these calculations shows that they do not depend significantly on what happens at distances that are much shorter than the Compton wave-length of the electron (about 3×10^{-12} m.), while other considerations which we shall discuss presently suggest that the failure of current theories should first become important around 10^{-20} m. The agreement of these calculations with experiment then constitutes an excellent verification of the current quantum theory in the domain in which all other indications suggest that it ought to be valid. However, it has also become clear that because this kind of experiment is so insensitive to the details of what happens in the domain of very short distances, it does not provide a very promising tool for investigating this domain.

On the other hand, experiments with particles of very high energy (of the order of 100 million electron volts or more) have led to a bewildering array of new phenomena, for which there is no adequate treatment in the existing theory. For as we have already pointed out in previous chapters, one discovers that the so-called "elementary particles" of physics, such as protons and neutrons, can now transform into each other. Moreover, many new particles, the positron, the neutrino, about ten different kinds of "mesons" and several new kinds of particles called hyperons have been discovered. No visible limit to this process of discovering new particles appears to be in sight as yet. And most of these new particles are unstable, having the ability to transform into each other, and to "decay" ultimately into neutrons, electrons, and protons. Besides, they can all be "created" in energetic collisions of other particles with nuclei. Moreover, a more accurate analysis of the data suggests that these new properties of matter become important only when particles approach within a distance of each other that is of the order of 10^{-20} m or less. Hence, in experiments carried out at the atomic level, practically no indication of these new properties is to be found.

It is evident, then, that the entire scheme by which the universe is regarded as made of certain kinds of "elementary particles" has demonstrated its inadequacy, and that some very different concept is needed here. Thus, when a similar instability and transformability of atoms in radioactive transformations was discovered half a century ago, it soon became evident that the chemical "elements" were not really elementary, being composed in fact, as was discovered later, of protons, electrons, and neutrons. Similarly, it seems reasonable to conclude that in the domain of very high energy experiments, we are disturbing the present-day "elementary" particles sufficiently so that their actual structure is beginning to manifest itself. According to the considerations that we have discussed previously, this structure should have a size of the order of 10^{-20} m.

It is easy to see that there are strong reasons for supposing a connection between the problem of the structure of the "elementary" particles and that of the infinities predicted by current theories. For if particles have a structure, this already implies that they occupy some space. And if they occupy space, then they will not be mathematical points, so that there will be no occasion for these infinities to arise. Just what the internal structure of these particles is we do not know as yet, but to find out is now our problem. Evidently, it must be something new relative to what is known so far. In the next section we shall discuss the indications that now exist regarding the nature of this structure.

We shall now discuss some of the principal advantages of the suggested new interpretation of the quantum theory over the usual interpretation in the guidance of research aimed at resolving this crisis.

First of all, let us recall that one of the principal problems now faced in this domain is that of treating the structure of an "elementary" particle, and of discovering what kinds of motions are taking place within this structure—motions that would help explain, perhaps, the "creation" and "destruction" of various kinds of particles, and their transformation into each other. In the usual interpretation of the quantum theory, it is extraordinarily difficult to consider this problem. For the insistence that one is not to be allowed to conceive of what is happening at this level means that one is restricted to making blind mathematical manipulations with the hope that somehow one of these manipulations will lead to a new and correct theory.

Secondly, the usual interpretation of the quantum theory implies a certain general mathematical and physical scheme which does not seem to lend itself very well to the notion that matter has new kinds of properties connected with an inner structure of the "elementary" particles. This general scheme, which was mentioned in [6.1] is the one involving purely linear equations for a wave function in a configuration space, "observables" obtained from linear operators, a purely probabilistic interpretation of the wave functions, etc. If one adopts this scheme, then the only mathematical possibilities left open seem to be the modification of current versions of the quantum theory, by alterations of the equations in such a way as somehow to cut out the contributions from short distances that lead to the infinities. Throughout the past twenty years, a great deal of intensive research has been devoted to attempts to do this in various ways (by cut-offs, finite distance operators, S-matrices, etc.), but none of these efforts has as yet shown any promise of leading to a consistent theory. These attempts have in general been guided by the expectation, commonly held by modern theoretical physicists, that in future theories the behaviour of things will be even less precisely definable than is possible in current theories. Of course, it cannot be proved at present that these expectations are definitely wrong. But the failure of the large number of efforts that have already been made in this direction would suggest that it may well be fruitful to try other lines of approach, especially considering that, as we have seen in the previous chapter, the restriction to the currently accepted line of approach is in any case not justifiable by any experimental or theoretical evidence coming from physics itself.

On the other hand, when we attack these problems within the framework of the new interpretation of the quantum theory, a large number of interesting new possibilities are seen to open up. First of all, the work is considerably facilitated by the fact that we can imagine what is happening, so that we can be led to new ideas not only by looking directly for new equations, but also by a related procedure of thinking in terms of concepts and models that will help to suggest new equations, which would be very unlikely to be suggested by mathematical methods alone. More important, however, is the fact that in terms of the notion of a sub-quantum mechanical level, we are enabled to consider a whole range of qualitatively new kinds of theories, approaching the usual form of the quantum theory only as approximations that hold in limiting cases. Moreover, there are a number of reasons, suggesting that the new features of such theories are likely to be relevant in the treatment of processes involving very high energies and very short distances. Some of these reasons are:

If there is a sub-quantum mechanical level, then, processes with very high energy and very high frequency may be faster than the processes taking place in the lower level. In such cases, the details of the lower level would become significant, and the current f ormulation of the quantum theory would break down.

For example, in our point of view, the "creation" of a particle, such as a meson, is conceived as a well-defined process taking place in the sub-quantum mechanical level. In this process, the field energy is concentrated in a certain region of space in discrete amounts, while the "destruction" of the particle is just the reverse process, in which the energy disperses and takes another form. In the quantum domain, the precise details of this process are not significant, and can therefore be ignored. This is in fact what is done in the current quantum theory which discusses the "creation" and "destruction" of particles as merely a kind of "popping" [6.1] in and out of being in a way that is simply not supposed to be subject to further description. With very fast high energy processes, however, the results may well depend on these details, and if this should be the case, the current quantum theory would not be adequate for the treatment of such processes.

REFERENCES

[6.1] Bohm, D. *Phys. Rev.* 1952, *85*, 166; Bohm, D. *Phys. Rev.* 1952, *85*, 180; Bohm, D. *Prog. Theor. Physics.* 1953, *9*, 273; Bohm, D.; Vigier, J. P. *Phys. Rev.* 1954, *96*, 208; Bohm, D.; Schiller, R.; Tiomno, J. *Supplemento al Nuovo Cimento* 1953, I Serie X, 48.

[6.2] Furth, R. *Zeits. f. Phys.* 1933, *81*, 143.

[6.3] de Broglie, L. *Compt. Rend.* 1926, *183*, 447; 1927, *185*, 380.

[6.4]] de Broglie, L. *The Revolution in Physics*; Routledge and Kegan Paul: London, 1954.

[6.5] Madelung, E. *Zeits. f. Phys.* 1926, *40* , 332.

[6.6] *Uspekhi. fizich. Nauk,* 1951, *45*; French translation in *Questions Scientifiques*; Editions de la Nouvelle Critique : Paris, 1952 ; Vol. 1. Blokhinzhev, D. J. *Grundlagen der Quantenmechanik ;* Deutscher Verlag der Wissenschaften: Berlin, 1953.

[6.7] de Broglie, L. *La Physique Quantique Resteratelle Indeterministe*; Gauthier-Villars: Paris, 1953.

[6.8] Fukada, M.; Miyamoto, K.; Tomanaga, S. *Prog. Theor. Physics* 1949, *4*, 47, 121.

[6.9] Schwinger, J. *Phys. Rev.* 1949, *74*, 749, 769; 1950, *80*, 440.

[6.10] Feynman, R. *Phys. Rev.* 1949, *75*, 486, 1736.

Chapter 7

PROCA EQUATION FOR HEAVY PHOTONS AND GRAVITONS

7.1. INTRODUCTION

In Chapter 4 we studied the application of the Proca thermal equation for the electron and nucleon gases. In this Chapter we formulated the description for thermal processes in heavy photon and graviton gases

In Chapter 4 the relativistic hyperbolic transport equation was developed:

$$\frac{1}{\upsilon^2}\frac{\partial^2 T}{\partial t^2}+\frac{m_0\gamma}{\hbar}\frac{\partial T}{\partial t}=\nabla^2 T. \tag{7.1}$$

In equation (7.1) υ is the velocity of heat waves, m_0 is the mass of heat carrier and γ – the Lorentz factor, $\gamma=\left(1-\frac{\upsilon^2}{c^2}\right)^{-1/2}$. As was shown Chapter 4 the heat energy (*heaton temperature*) T_h can be defined as follows:

$$T_h = m_0\gamma\upsilon^2. \tag{7.2}$$

Considering that υ, the thermal wave velocity equals [7.1]

$$\upsilon = \alpha c \tag{7.3}$$

where α is the coupling constant for the interactions which generate the *thermal wave* (α = 1/137 and α = 0.15 for electromagnetic and strong forces respectively). The *heaton temperature* is equal to

$$T_h=\frac{m_0\alpha^2 c^2}{\sqrt{1-\alpha^2}}. \tag{7.4}$$

Based on equation (7.4) one concludes that the *heaton temperature* is a linear function of the mass m_0 of the heat carrier. It is interesting to observe that the proportionality of T_h and the heat carrier mass m_0 was observed for the first time in ultrahigh energy heavy ion reactions measured at CERN [7.2]. In monograph [7.2] it was shown that the temperature of pions, kaons and protons produced in Pb+Pb, S+S reactions are proportional to the mass of particles. In this monograph the damped thermal wave equation was developed:

$$\frac{1}{\upsilon^2}\frac{\partial^2 T}{\partial t^2}+\frac{m}{\hbar}\frac{\partial T}{\partial t}+\frac{2Vm}{\hbar^2}T-\nabla^2 T=0. \tag{7.5}$$

The relativistic generalization of equation (7.1) is quite obvious:

$$\frac{1}{\upsilon^2}\frac{\partial^2 T}{\partial t^2}+\frac{m_0\gamma}{\hbar}\frac{\partial T}{\partial t}+\frac{2Vm_0\gamma}{\hbar^2}T-\nabla^2 T=0. \tag{7.6}$$

It is worthwhile noting that in order to obtain a non-relativistic equation we put $\gamma = 1$.

When the external force is present $F(x,t)$ the forced damped heat transport is obtained instead of equation (7.6) (in one dimensional case):

$$\frac{1}{\upsilon^2}\frac{\partial^2 T}{\partial t^2}+\frac{m_0\gamma}{\hbar}\frac{\partial T}{\partial t}+\frac{2Vm_0\gamma}{\hbar^2}T-\frac{\partial^2 T}{\partial x^2}=F(x,t). \tag{7.7}$$

The hyperbolic relativistic quantum heat transport equation, (7.7), describes the forced motion of heat carriers which undergo scattering ($\frac{m_0\gamma}{\hbar}\frac{\partial T}{\partial t}$ term) and are influenced by the potential term ($\frac{2Vm_0\gamma}{\hbar^2}T$).

Equation (7.7) can be written as

$$\begin{gathered}\left(\bar{\Box}^2+\frac{2Vm_0\gamma}{\hbar^2}\right)T+\frac{m_0\gamma}{\hbar}\frac{\partial T}{\partial t}=F(x,t),\\ \bar{\Box}^2=\frac{1}{\upsilon^2}\frac{\partial^2}{\partial t^2}-\frac{\partial^2}{\partial x^2}.\end{gathered} \tag{7.8}$$

We seek the solution of equation (7.7) in the form

$$T(x,t)=e^{-t/2\tau}u(x,t), \tag{7.9}$$

where $\tau = \hbar/(m\upsilon^2)$ is the relaxation time. After substituting equation (7.9) in equation (7.8) we obtain a new equation

$$\left(\bar{\square}^2 + q^2\right) u(x,t) = e^{t/2\tau} F(x,t) \tag{7.10}$$

and

$$q^2 = \frac{2Vm}{\hbar^2} - \left(\frac{mv}{2\hbar}\right)^2 \tag{7.11}$$

$$m = m_0 \gamma \tag{7.12}$$

In free space i.e. when $F(x,t) \rightarrow 0$ equation (7.11) reduces to

$$\left(\bar{\square}^2 + q^2\right) u(x,t) = 0 \tag{7.13}$$

which is essentially the free *Proca* equation.

The *Proca* equation describes the interaction of the laser pulse with the matter. As was shown in Chapter 4 the quantization of the temperature field leads to the *heatons* – quanta of thermal energy with a mass $m_h = \hbar / \tau v_h^2$ [7.1], where τ is the relaxation time and v_h is the finite velocity for heat propagation. For $v_h \rightarrow \infty$, i.e. for $c \rightarrow \infty$, $m_o \rightarrow 0$ it can be concluded that in non-relativistic approximation (c = infinite) the *Proca* equation is the diffusion equation for massless photons and heatons.

It is well known that the mass of the photon in vacuum must be non-zero due to Heisenberg's uncertainty principle[7.3]. There is also an ongoing discussion in the literature if it is possible that the graviton has a non-zero mass as well due to the measurement of the cosmological constant in the universe[7.4 – 7.6]. Both Heisenberg's limit and the cosmological constant lead respectively to a photon and a graviton mass of about 10^{-69} kg. This is obviously very small and it is therefore believed that it has - negligible consequences. We will show that this is actually not the case and leads to fundamental new insights both for classical and quantum matter. An immediate consequence is that Maxwell equations transform into Proca type equations leading us to the conclusion that for the electromagnetic and gravitational interaction gauge invariance only applies to a certain approximation in free space. The consequence of this result will be discussed in the following. Especially the new treatments required for the graviton have far reaching consequences that can be experimentally assessed.

7.2. Proca Equations and the Photon Mass

As the photon's mass is non-zero, the usual Maxwell equations for electromagnetism transform into the well known Proca equations with additional terms due to the finite photon wavelength

$$div\vec{E}=\frac{\rho}{\varepsilon_0}-\left(\frac{m_\gamma c}{\hbar}\right)^2\cdot\varphi,$$

$$div\vec{B}=0,$$

$$rot\vec{E}=-\frac{\partial\vec{B}}{\partial t}, \tag{7.14}$$

$$rot\vec{B}=\mu_0\rho\vec{\upsilon}+\frac{1}{c^2}\frac{\partial\vec{E}}{\partial t}-\left(\frac{m_\gamma c}{\hbar}\right)^2\cdot\vec{A}.$$

The usual effect attributed to a finite photon mass is its consequence on the strength of electromagnetic forces over distances. However, taking the curl of the 4th equation reveals another feature, which was first assessed by the authors in the framework of superconductivity [7.7]

$$B=B_0\cdot e^{-\frac{x}{\lambda_\gamma}}+2\omega\rho\mu_0\lambda_\gamma^2. \tag{7.15}$$

The first part of Eq. (7.15) is the Yukawa-type exponential decay of the magnetic field, and the second part shows that a magnetic field will be generated due to the rotation of a charge density ρ. In quantum field theory, superconductivity is explained via a large photon mass as a consequence of gauge symmetry breaking and the Higgs mechanism. The photon wavelength is then interpreted as the London penetration depth and leads to a Photon mass about 1/1000 of the electron mass. This then leads to

$$B=B_0\cdot e^{-\frac{x}{\lambda_\gamma}}-2\frac{m^*}{e^*}\omega, \tag{7.16}$$

where the first term is called the Meissner-Ochsenfeld effect (shielding of electromagnetic fields entering the superconductor) and the second term is known as the London moment (minus sign is due to the fact that the Cooper-pairs lag behind the rotation) with m^* and e^* as the Cooper-pair's mass and charge (called Becker's argument [7.8]). The magnetic field produced by a rotating superconductor as a consequence of its large photon mass was experimentally measured outside of the quantum condensate where the photon mass is believed to be close to zero[7.9, 7.10].

Knowing the effect from a large photon mass in superconductivity, what is the effect of a non-zero photon mass in normal matter? Contrary to the case of superconductivity where the photon mass in comparable to the mass on an electron, in normal matter, the photon mass is believed to be close to zero. The problem of the Proca equations is easily shown: The "Meissner" part becomes important only for large photon masses (which is not the case in normal matter) - but the "London Moment" part becomes important for very small photon masses. By taking the presently accepted experimental limit on the photon mass [7.3] ($m_\gamma < 10^{-52}$ kg), we can write the second part as

$$B > \omega\rho \cdot 2.8 \times 10^{13}. \tag{7.17}$$

This would mean that a charge density ρ rotating at an angular velocity *co* should produce huge magnetic fields. Obviously, this is not the case leading to a paradox for normal matter. Therefore the value of the photon mass for matter containing a charge density must be different from the one in free space.

In fact, there is only one possible choice for the photon mass. Let's consider the case of a single electron moving in a magnetic field. It will then perform a precession movement according to Larmor's theorem ω=-$(q/2m)B$. We can then solve Eq. (7.15) for the photon mass expressing it as

$$\frac{1}{\lambda_\gamma^2} = \left(\frac{m_\gamma c^2}{\hbar}\right)^2 = -\frac{q}{m}\mu_0\rho. \tag{7.18}$$

We find that the photon mass inside normal matter must be defined over the charge density and the charge-to-mass ratio observed. Of course, as the Larmor theorem describes pseudo forces in rotating reference frames, the photon mass in Eq. (7.18) is not a real mass but can be interpreted as an equivalent photon mass inside the material necessary to comply with Newton's mechanics. Note that this equivalent photon mass is then (due to the negative sign in the Larmor theorem) always a complex value independent of the sign of charge. This is opposite to the case of superconductivity where the photon mass due to the Higgs mechanism indeed has a real value.

Only for neutral matter or vacuum, the photon mass can be therefore given by the limit obtained via Heisenberg's uncertainty principle. As nearly all matter in the universe can be considered neutral, this consequence may be of minor importance. However, in case of gravity and the graviton, the consequences are far reaching as matter is never neutral in a gravitational sense.

7.3. PROCA EQUATIONS AND GRAVIPHOTON MASS

In the weak field approximation, gravity can be written similar to a Maxweilian structure forming the so-called Einstein-Maxwell equations. The quantization of the Maxweilian theory of gravity would lead to a spin one boson as a mediator of gravitoelectromagnetic fields. This represents a major problem with respect to the theory of general relativity, which only predicts quadrupolar gravitational waves associated with spin two gravitons. This is the reason why the linear approximation of Einstein field equations is taken as being only an approximation to the complete theory, which cannot be used to investigate radiative processes. Recent experimental results on the gravitomagnetic London moment [7.11] tend to demonstrate that gravitational dipolar type radiation associated with the Einstein-Maxwell equations is real. This implies that Maxweilian gravity is not only an approximation to the complete theory, but may indeed reveal a new aspect of gravitational phenomena associated with a vectorial spin 1 gravitational boson, which we might call the graviphoton. As an example, a fully relativistic modified theory of gravity called scalar vector tensor gravity

[7.12] would duly take into account this new side of gravity. In the following we will therefore use the term graviphoton for studying the Proca type character of gravity and its consequences on coherent matter.

The field equations for massive linearized gravity are given by [7.13]:

$$
\begin{aligned}
div\vec{g} &= \frac{\rho_m}{\varepsilon_g} - \left(\frac{m_g c}{\hbar}\right)^2 \cdot \varphi_g, \\
div\vec{B}_g &= 0, \\
rot\vec{g} &= -\frac{\partial \vec{B}_g}{\partial t}, \\
rot\vec{B}_g &= \mu_{0g}\rho_m \vec{\upsilon} + \frac{1}{c^2}\frac{\partial \vec{g}}{\partial t} - \left(\frac{m_g c}{\hbar}\right)^2 \cdot \vec{A}_g,
\end{aligned}
\tag{7.19}
$$

where g is the gravitoelecthc and B_g the gravitomagnetic field. Applying again a curl on the 4th Proca equation leads to

$$
B_g = B_{0g} \cdot e^{-\frac{x}{\lambda_\gamma}} - 2\omega\rho_m\mu_{0g}\lambda_g^2, \tag{7.20}
$$

where ρ_m is now the mass density and μ_{0g} the gravitomagnetic permeability (note the different sign with respect to Eq. (7.15)). Similar to electromagnetism, we obtain a Meissner and a London moment part for the gravitomagnetic field generated by matter [7.7].

7.4. Consequences of Local Graviton Mass

Applying a non-zero graviphoton mass to Eq. (7.20) leads again to huge gravitomagnetic fields for rotating mass densities which are not observed. Again we find the solution in the gravitational analog to the Larmor theorem. Locally, the principle of equivalence must be fulfilled for any type of matter. That means that local accelerations must be equivalent to gravitational fields and a body can not distinguish between being in a rotating reference frame or being subjected to a gravitomagnetic field ($B = -2\omega$).

By choosing the graviphoton wavelength proportional to the local density of matter similar to Eq. (7.18),

$$
\frac{1}{\lambda_g^2} = \left(\frac{m_g c^2}{\hbar}\right)^2 = \mu_{0g}\rho_m, \tag{7.21}
$$

we find the solution. Note that due to the different signs in the Einstein-Proca equations, we find that the equivalent graviphoton mass inside normal matter is a real number to comply with the Larmor theorem in normal matter. Inserting Eq. (7.21) into our Proca equations for gravity in Eq. (7.19), we find by performing another grad operator on the 1st equation and another curl on the 4th equation

$$g = -a, \qquad B_g = -2\omega, \tag{7.22}$$

which is nothing else as the formulation of the equivalence principle and the gravitomagnetic Larmor theorem [7.14]. The equivalent graviphoton mass in Eq. (7.21) therefore describes the inertial properties of matter, in accelerated reference frames. This is a very fundamental result and new insight into the foundations of mechanics.

A similar graviphoton mass (up to a factor 3/16) was already found by Argyris [7.13] by solving Einstein's equations in the conformally flat case. He linked it with the average mass of the universe. However, as we have shown, this result is valid locally for all matter.

It is interesting to note that if we take the case of no local sources (ρ_m = 0), the graviphoton mass will be zero, and we will find, by solving the weak field equations in the transverse gauge, the "classical" freely propagating degrees of freedom of gravitational waves associated with a massless spin 2 graviton [7.15]. However, in the case of local sources, a spin-1 graviphoton will appear.

7.5. Application to Coherent Quantum Matter

Perhaps that most important consequence of the local graviphoton mass is its relation to superconductivity. As we wrote already in the introduction of this paper, the application of Proca equations to superconductivity are well established. In a superconductor, we have now a ratio between matter being in normal and in a condensed (coherent) state. So we have two sets of Proca equations, one which deals with the overall mass and one with its condensated subset.

The coherent part of a given material (e.g. the Cooper-pair fluid) is also described by its own set of Proca equations similar to the ones in Eq. (7.19) but with one important difference: Instead of the ordinary mass density ρ_m we have to take the Cooper-pair mass density ρ_m^*. By taking the curl of the fourth Proca equation [7.7], we arrive at

$$B_g = B_{0g} \cdot e^{-\frac{x}{\lambda_\gamma}} + 2\omega\rho_m^*\mu_{0g}\lambda_g^2, \tag{7.23}$$

where we had to introduce Becker's argument [7.8] that the Cooper-pairs are lagging behind the lattice so that the current is flowing in the opposite direction of ω. This gives the right sign with respect to our experimental observation [7.11]. A similar argument was introduced for the classical London moment as also here the sign change was observed accordingly. The first part (Meissner part) is not different from our previous assessment for normal matter, but the second part changes to

$$B_g = 2\omega \frac{\overset{*}{\rho}_m}{\rho_m}, \tag{7.24}$$

due to the fact that the graviphoton mass depends on all matter in the material, not just the coherent part. That is why the densities do not cut any more and we arrive at an additional field for coherent matter in addition to the gravitational Larmor theorem. As an alternative to Becker's arguments [7.8] on the sign change in superconductors, one could also switch between real and imaginary values for the equivalent photon and graviphoton when going from the normal to the coherent state of matter.

Similarly, by taking the gradient of the first Proca equation in Eq. (7.19), considering the case of an homogeneous gravitational field, and using the fact that the gradient of a density of energy is equal to a density of force, $\nabla\left(\overset{*}{\rho}_m c^2\right) = \overset{*}{\rho}_m \vec{a}$ we obtain:

$$\vec{g} = \vec{a}\overset{*}{\rho}_m \mu_{0g} \lambda_g^2, \tag{7.25}$$

which transforms using Eq.(7.22), into:

$$\vec{g} = \frac{\overset{*}{\rho}_m}{\rho_m} \vec{a}, \tag{7.26}$$

where $\vec{a}$ is the total net acceleration to which the superconductor is submitted.

Comparing Eqs. (7.24) and (7.26) to Eq. (7.22) we conclude that the presence of Cooper-pairs inside the superconductor leads to a deviation from the equivalence principle and from the classical gravitational Larmor theorem. A rigid reference frame mixed with non-coherent and coherent matter is not equivalent to a rigid reference frame made of normal matter alone, with respect to its inertial and gravitational properties. However, in the case of a Bose-Einstein condensate where we have only coherent matter, Eq. (7.26) transforms into the usual expression for normal matter and the equivalence principle is again conserved.

An important feature is that these fields (Eqs. (7.24) and (7.26)) however, should be also present *outside* the superconductor, contrary to the classical inertial behaviour. The gravitomagnetic Larmor theorem for normal matter describes the inertial forces in an accelerated reference frame. This leads to so-called pseudo forces which are only present inside the material which is rotating. The difference to quantum materials is that in this case, the integral of the canonical momentum is quantised. Let's consider a'superconducting ring. The integral of the full canonical momentum of the Cooper-pairs including gravitational fields is given by

$$\oint \vec{p}_s \cdot d\vec{l} = \oint \left(m^* \vec{v}_s + e^* \vec{A} + m^* \vec{A}_g\right) \cdot d\vec{l} = \frac{nh}{2}. \tag{7.27}$$

If the ring is thicker than the London penetration depth, then the integral can be set to zero. Solving for the case where the superconductor is at an angular velocity ω, we get

$$\vec{B} = -2\frac{m}{e}\vec{\omega} - \frac{m}{e}\vec{B}_g. \tag{7.28}$$

These magnetic and gravitomagnetic fields are also present inside the superconducting ring. The first part is the classical London moment, with its origin is due to the photon mass, and the second part is its analog gravitomagnetic London moment, which will produce an additional field overlapping the classical London moment. According to Eq. (7.24), depending on the superconductor's bulk and Cooper-pair density, the magnetic field should be higher than classically expected. Indeed, that has been measured without apparent solution throughout the literature.

The authors [7.16] already conjectured such a field to explain a reported disagreement between the theoretical and experimental Cooper-pair mass [7.7, 7.17, 7.18]. Tate et al. [7.9, 7.10] used a sensitive London moment measurement to determine the Cooper-pair mass in Niobium. This mass was found to be larger ($m^*/2m_e$ = 1.000084(21)) than the theoretically expected value ($m^*/2m_e$ = 0.999992). As we pointed out earlier in our conjecture, this mass difference opens up the room for large gravitomagnetic fields following the quantised canonical momentum in Eq. (7.25). In order to correct Tate's result, we need a gravitomagnetic field of

$$B = 2\frac{m^*_{\text{experimental}} - m^*_{theoretical}}{\mathrm{m}^*_{\text{theo}}}\omega = 2\frac{\Delta m^*}{m^*}\omega, \tag{7.29}$$

where Δm^* is the difference between the experimental and theoretical Cooper-pair mass to find back the Cooper-pair mass predicted by quantum theory.

The local graviton mass now establishes the reason why such a gravitomagnetic field has to be there. Comparing Eqs. (7.24) and (7.29), we can identify

$$\frac{\Delta m^*}{m^*} = \frac{\rho^*_m}{\rho_m}. \tag{7.30}$$

Taking Tate's values $\left(\frac{\Delta m^*}{m^*} = 9.2\times10^{-5}\right)$ and the Niobium bulk and Cooper-pair mass density $\left(\frac{\rho^*}{\rho} = 3.95\times10^{-6}\right)$, we see that these values are a factor of 23 away. One has to take into account that Tate's measurement is up to now the only precision experiment and, even more important, other relativistic correction terms need to be added to the theoretical Cooper-pair mass, which will make Δm^* smaller and Eq. (7.30) match better.

This is a very important result as we have for the first time not only a conjecture to explain Tate's anomaly, but also a good reason why a rotating superconductor should produce a gravitomagnetic field which is larger than classical predictions from ordinary rotating matter. The reason is the local (equivalent) graviphoton mass.

First measurements show that this gravitomagnetic field indeed exists with a measurement in between our Cooper-pair density ratio and the one derived from Tate's measurements $\left(\frac{\Delta m^*}{m^*}_{measured} \cong 2.6\times 10^{-5}\right)$. This adds strong confidence in our theoretical approach and its consequences.

It was shown that non-zero values for the graviphoton leads to huge gravitomagnetic fields around rotating mass densities, which are not observed. The solution to the problem is found by an equivalent graviphoton mass which depends on the local mass density leading to the correct inertial forces in rotating reference frames. That can be understood as a foundation of basic mechanics. This solution, derived from Einstein and Proca equations, has important consequences such as for the case of normal matter, the Proca equations now lead to the gravitomagnetic Larmor theorem and not to unobserved huge gravitomagnetic fields, the prediction of a gravitomagnetic London moment (observed experimentally) in rotating superconductors, that can solve the Cooper-pair mass anomaly reported by Tate, among many others. Similar results have also been outlined for the Photon mass. By obeying Larmor's theorem, we find that in classical matter the equivalent photon has a complex and the graviphoton a real value. In coherent matter we suggest the hypothesis that it is exactly the other way round, which solves the sign change problems associated to the classical and gravitomagnetic London moment as an alternative to the usual Becker argument.

Let us formulate the Cauchy problem for the heat transport in heavy photons and gravitons [7.2].

For the initial *Cauchy* condition:

$$u(x,0) = f(x), \qquad u_t(x,0) = g(x) \tag{7.31}$$

the solution of the *Proca* equation has the form (for $q^2 > 0$) [7.2]

$$\begin{aligned} u(x,t) &= \frac{f(x-\upsilon t)+f(x+\upsilon t)}{2} \\ &+ \frac{1}{2\upsilon}\int_{x-\upsilon t}^{x+\upsilon t} g(\varsigma)J_0\left[\sqrt{q^2\left(\upsilon^2t^2-(x-\varsigma)^2\right)}\right]d\varsigma \\ &- \frac{\sqrt{q}\upsilon t}{2}\int_{x-\upsilon t}^{x+\upsilon t} f(\varsigma)\frac{J_1\left[\sqrt{q^2\left(\upsilon^2t^2-(x-\varsigma)^2\right)}\right]}{\sqrt{\upsilon^2t^2-(x-\varsigma)^2}}d\varsigma \\ &+ \frac{1}{2\upsilon}\int_0^t\int_{x-\upsilon(t-t')}^{x+\upsilon(t-t')} G(\varsigma,t')J_0\left[\sqrt{q^2\left(\upsilon^2(t-t')^2-(x-\varsigma)^2\right)}\right]dt'd\varsigma. \end{aligned} \tag{7.32}$$

where $G = e^{t/2\tau}F(x,t)$.

When $q^2 < 0$ solution of *Proca* equation has the form:

$$u(x,t)=\frac{f(x-\upsilon t)+f(x+\upsilon t)}{2}$$
$$+\frac{1}{2\upsilon}\int_{x-\upsilon t}^{x+\upsilon t} g(\varsigma)I_0\left[\sqrt{-q^2\left(\upsilon^2t^2-(x-\varsigma)^2\right)}\right]d\varsigma$$
$$-\frac{\sqrt{-q^2}\,\upsilon t}{2}\int_{x-\upsilon t}^{x+\upsilon t} f(\varsigma)\frac{I_1\left[\sqrt{-q^2\left(\upsilon^2t^2-(x-\varsigma)^2\right)}\right]}{\sqrt{\upsilon^2t^2-(x-\varsigma)^2}}d\varsigma \tag{7.33}$$
$$+\frac{1}{2\upsilon}\int_0^t\int_{x-\upsilon(t-t')}^{x+\upsilon(t-t')} G(\varsigma,t')I_0\left[\sqrt{-q^2\left(\upsilon^2(t-t')^2-(x-\varsigma)^2\right)}\right]dt'd\varsigma.$$

When $q^2 = 0$ equation (7.13) is the forced thermal equation

$$\frac{1}{\upsilon^2}\frac{\partial^2 u}{\partial t^2}-\frac{\partial^2 u}{\partial x^2}=G(x,t). \tag{7.34}$$

On the other hand one can say that equation (7.13) is distortion-less hyperbolic equation. The condition $q^2 = 0$ can be rewritten as:

$$V\tau=\frac{\hbar}{8} \tag{7.35}$$

The equation (7.35) is the analogous to the Heisenberg uncertainty relation. Considering equation (7.35) can be written as:

$$V=\frac{T_h}{8}, \qquad V<T_h. \tag{7.36}$$

It can be stated that distortion-less waves can be generated only if $T_h > V$. For $T_h < V$, i.e. when the "Heisenberg rule" is broken, the shape of the thermal waves is changed.

In this chapter we developed the relativistic thermal transport equation for heavy photon and graviton gases. It is shown that the equation obtained is the *Proca* equation, well known in relativistic electrodynamics.

References

[7.1] Pelc, M. *Non-Fourier description of the laser pulse matter interaction*; Dissertation; University of Marie Curie – Sklodowska, Physics Dept.: Lublin, Poland, 2008.

[7.2]Kozlowski, M.; Marciak – Kozlowska, J. *Thermal Processes Using Attosecond Laser Pulses*, Optical Science 121; Springer: New York, NY, 2006.

[7.3]. Tu, L.; Luo, J.; Gillies, J. T. *Rep. Prog. Phys.* 2005, *68*, 77-130.

[7.4]. Spergel, D. N.; et al. *Astrophy. J. Suppl.* 2003, *148*, 175.

[7.5]. Novello, M.; Neves, R. P. *Class. Quantum Grav.* 2003, *20*, L67-L73.
[7.6]. Liao, L. (2004). On the gravitational wave in de Sitter space – time. arXiv/gr-qc/0411122.
[7.7]. de Matos, C. J.; Tajmar, M. *Physica* 2005, *C432*, 167-172.
[7.8]. Becker, R.; Heller, G.; Sauter, F. Z. *Physik* 1933, *S5,* 772-787.
[7.9]. Tate, J.; et al. *Phys. Rev. Lett.* 1989, *62(8)*, 845-848.
[7.10] Tate, J.; et al. *Phys. Rev. B* 1990, *42*(*13*), 7885-7893.
[7.11] Tajmar, M.; Plecsu, R. (2006). Experimental detection of the gravitational London Moment. arXiv:0603033.
[7.12] Moffat, J. W. (2005). Scalar tensor – vector gravity theory. arXiv:0506021.
[7.13] Argyris, J.; Ciubotariu, C. *Aust. J. Phys.* 1997, *50*, 879-891.
[7.14] Mashhoon, B. *Phys. Lett. A* 1993, *173*, 347-354.
[7.15] Carroll, S. *Spacetime and Geometry: An Introduction to General Relativity*; Addison Wesley: New York, NY, 2003; 293-300.
[7.16] Tajmar, M.; de Matos, C. J. (2006). Local photon and graviton mass and its consequences. arXiv:0603032.
[7.17] Tajmar, M.; de Matos, C. J. *Physica C* 2003, *385*(*4*), 551-554.
[7.18] Tajmar, M.; de Matos, C. J. *Physica C* 2005, *420*(*1-2*), 56-60.

Appendix A. Laser Induced Transport Phenomena in Graphene

In the description of the evolution of any physical system, it is mandatory to evaluate as accurately as possible the order of magnitude of different characteristic time scales, since their relationship with the time scale of observation (the time during which we assume our description of the system is valid) will determine along with the relevant equation pattern.

The advent of attosecond laser pulses opens the new field of investigation of the quantum phenomena. As all measured relaxation times are much longer the attosecond laser pulses "observe" the generic quantum nature of the phenomena, not averaged over time.

In monograph [A.1] the theoretical framework for transport processes generated by attosecond laser pulses was formulated. It was shown that the master equation is the hyperbolic transport equation:

$$\frac{1}{\upsilon^2}\frac{\partial^2 T(\vec{r},t)}{\partial t^2} - \Delta T(\vec{r},t) + q^2 T(\vec{r},t) = 0, \tag{A.1}$$

$$q^2 = \frac{2Vm}{\hbar^2} - \left(\frac{m\upsilon}{2\hbar}\right)^2, \tag{A.2}$$

$$\upsilon = \alpha c,$$

where $\alpha = 1/137$ is the electromagnetic coupling constant and c is the light velocity, m - heat carrier mass.

Depending on the sign of the q the equation is the Heaviside equation ($q^2 < 0$) or Klein - Gordon equation ($q^2 > 0$). For $q^2 = 0$ Eq. (A.1) is the wave equation which describes the ballistic, quasi-free propagation of carriers.

Recent measurement of the electrical and thermal properties of graphenes shed new light on the application of the equation (A.1) to the investigation of the transport phenomena.

As an example we invoke the ballistic transport in graphene which as can be seen from Eq. (A1) is the result of the $q^2 = 0$, moreover it occurs that the velocity of fermions in graphene is of the order of 10^6 m/s, which agrees with $\upsilon = \alpha c$ [A.2 – A.4].

The aim of this Appendix is the solution of the equation (A.1) for the Klein - Gordon and Heaviside branches.

1. The Model Equations

Given the initial value problem for hyperbolic thermal equation

$$u_{tt} - \gamma^2 u_{xx} + c^2 u = F(x,t) \tag{A.3}$$

$$-\infty < x < \infty, \qquad t > 0$$

with the initial conditions

$$u(x,0) = f(x), \qquad u_t(x,0) = g(x), \tag{A.4}$$

$$-\infty < x < \infty,$$

the solution at an arbitrary point (ξ, τ) is given by

$$u(\xi,\tau) = \int_0^T \int_{-\infty}^{\infty} FK dx dt + \int_{-\infty}^{\infty} \left[gK(x,0;\xi,\tau) - f \frac{\partial K}{\partial t}(x,0;\xi,\tau) \right] dx, \tag{A.5}$$

where $\tau < T$ and $K(x,0;\xi,\tau)$ is the solution of the equation

$$\frac{\partial^2 K}{\partial t^2} - \gamma^2 \frac{\partial^2 K}{\partial x^2} + c^2 K = \delta(x-\xi)\delta(t-\tau). \tag{A.6}$$

As can be seen from equation (A.6) $K(x,0;\xi,\tau)$ is the Green function for the equation (A.3) and fulfils the formula

$$K(x,t;\xi,\tau) = \frac{1}{2\gamma} J_0 \left[\frac{c}{\gamma} \sqrt{\gamma^2 (t-\tau)^2 - (x-\xi)^2} \right] \qquad \text{for } |x-\xi| < \gamma(t-\tau)$$

$$K(x,t;\xi,\tau) = 0 \qquad \text{for } |x-\xi| > \gamma(t-\tau) \tag{A.7}$$

and $J_0(y)$ is the Bessel function of the order 0.

Introducing the Heaviside step function $H(z)$ we express formulae (A.7) as

$$K(x,t;\xi,\tau)=\frac{1}{2\gamma}J_0\left[\frac{c}{\gamma}\sqrt{\gamma^2(t-\tau)^2-(x-\xi)^2}\right]*H[x-\gamma t-(\xi-\gamma\tau)]H[\xi+\gamma\tau-(x+\gamma t)] \quad (A.8)$$

Since $J_0(0) = 1$ we see that (A.8) reduces to the Green function for the one dimensional wave equation if we set $c = 0$.

We now show that $K(x,0;\xi,\tau)$ satisfies (A.6). To that aim we set

$$K = J_0\hat{K}, \quad (A.9)$$

where $\hat{K}$ is the Green function for the wave equation and satisfies

$$\frac{\partial^2\hat{K}}{\partial t^2}-c^2\frac{\partial^2\hat{K}}{\partial x^2}=\delta(x-\xi)\delta(t-\tau) \quad (A.10)$$

with c replaced by γ. Then

$$\begin{aligned}\frac{\partial^2 K}{\partial t^2}-\gamma^2\frac{\partial^2 K}{\partial x^2}+c^2K &= \left\{\frac{\partial^2 J_0}{\partial t^2}-\gamma^2\frac{\partial^2 J_0}{\partial x^2}+c^2J_0\right\}\hat{K}\\ &+2\left\{\frac{\partial J_0}{\partial t}\frac{\partial\hat{K}}{\partial t}-\gamma^2\frac{\partial J_0}{\partial x}\frac{\partial\hat{K}}{\partial x}\right\}\\ &+\left\{\frac{\partial^2\hat{K}}{\partial t^2}-\gamma^2\frac{\partial^2\hat{K}}{\partial x^2}\right\}J_0\end{aligned} \quad (A.11)$$

For $J_0\left[\frac{c}{\gamma}\sqrt{\gamma^2(t-\tau)^2-(x-\xi)^2}\right]=1$ the first term in (A.11) is equal zero. Also we have

$$\begin{aligned}2\left\{\frac{\partial J_0}{\partial t}\frac{\partial\hat{K}}{\partial t}-\gamma^2\frac{\partial J_0}{\partial x}\frac{\partial\hat{K}}{\partial x}\right\}&=\frac{cJ_0'\left[\left(\frac{c}{\gamma}\right)\sqrt{\gamma^2(t-\tau)^2-(x-\xi)^2}\right]}{\sqrt{\gamma^2(t-\tau)^2-(x-\xi)^2}}*\\ &*\left\{\begin{matrix}[-\gamma(t-\tau)+(x-\xi)]\delta[x-\gamma t-(\xi-\gamma t)]H[\xi+\gamma\tau-(x-\gamma t)]+\\ [-\gamma(t-\tau)-(x-\xi)]H[x-\gamma t-(\xi-\gamma t)]\delta[\xi+\gamma\tau-(x+\gamma t)]\end{matrix}\right\}.\end{aligned} \quad (A.12)$$

The expression (A.12) vanishes since

$$\begin{aligned}&[-\gamma(t-\tau)+(x-\xi)]\delta[x-\gamma t-(\xi-\gamma t)]=\\&[-\gamma(t-\tau)+(x-\xi)]\delta[-\gamma(t-\tau)+(x-\xi)]=0.\end{aligned} \tag{A.13}$$

$$\begin{aligned}&[-\gamma(t-\tau)-(x-\xi)]\delta[\xi+\gamma\tau-(x+\gamma t)]=\\&[-\gamma(t-\tau)-(x-\xi)]\delta[-\gamma(t-\tau)-(x-\xi)]=0.\end{aligned} \tag{A.14}$$

on using $x\delta(x)=0$. Finally we have

$$\begin{aligned}\left\{\frac{\partial^2\hat{K}}{\partial t^2}-\gamma^2\frac{\partial^2\hat{K}}{\partial x^2}\right\}J_0 &= \delta(x-\xi)\delta(t-\tau)J_0\left[\tfrac{c}{\gamma}\sqrt{\gamma^2(t-\tau)^2-(x-\xi)^2}\right]=\\ &=\delta(x-\xi)\delta(t-\tau)J_0(0)=\delta(x-\xi)\delta(t-\tau).\end{aligned} \tag{A15}$$

It may be noted that with $c=i\hat{c}$ (where $i=\sqrt{-1}$) in the equation (A.3) we obtain Heaviside equation

$$\frac{\partial^2 K}{\partial t^2}-\gamma^2\frac{\partial^2 K}{\partial x^2}-\hat{c}^2K=\delta(x-\xi)\delta(t-\tau). \tag{A16}$$

Since $J_0(ix)=I_0(x)$, the modified Bessel function of zero order we obtain for K in place (A.8)

$$K(x,t;\xi,\tau)=\frac{1}{2\gamma}I_0\left[\frac{\hat{c}}{\gamma}\sqrt{\gamma^2(t-\tau)^2-(x-\xi)^2}\right]*H[x-\gamma t-(\xi-\gamma\tau)]H[\xi+\gamma\tau-(x+\gamma t)] \tag{A.17}$$

is the Green function.

In the double integral (A.5) we have $K=0$ for $t>\tau$ so that the limit in the t integral extends only up to τ. Also from (A.7) we conclude that K vanishes unless $|x-\xi|<\gamma(\tau-t)$ and this is equivalent to

$$\xi-\gamma(\tau-t)<x<\xi+\gamma(\tau-t). \tag{A.18}$$

Therefore we obtain

$$\int_0^T\int_{-\infty}^{\infty}FKdxdt=\frac{1}{2\gamma}\int_0^T\int_{\xi-\gamma(\tau-t)}^{\xi+\gamma(\tau-t)}F(x,t)J_0\left[\frac{c}{\gamma}\sqrt{\gamma^2(t-\tau)^2-(x-\xi)^2}\right]dxdt. \tag{A.19}$$

Further we have

$$K(x,0;\xi,\tau)=\frac{1}{2\gamma}J_0\left[\frac{c}{\gamma}\sqrt{\gamma^2\tau^2-(x-\xi)^2}\right]*H[x-(\xi-\gamma\tau)]H[\xi+\gamma\tau-x] \quad (A.20)$$

so that

$$\int_{-\infty}^{\infty}g(x)K(x,0;\xi,\tau)dx=\frac{1}{2\gamma}\int_{\xi-\gamma\tau}^{\xi+\gamma\tau}g(x)J_0\left[\frac{c}{\gamma}\sqrt{\gamma^2\tau^2-(x-\xi)^2}\right]dx. \quad (A.21)$$

Since the product of the Heaviside function vanishes outside the interval $(\xi-\gamma\tau,\xi+\gamma\tau)$. Finally

$$\begin{aligned}\partial K(x,0;\xi,\tau)=&-\frac{c\tau}{2}\frac{J_0'\left[\frac{c}{\gamma}\sqrt{\gamma^2\tau^2-(x-\xi)^2}\right]}{\sqrt{\gamma^2\tau^2-(x-\xi)^2}}H[x-(\xi-\gamma\tau)]H[\xi+\gamma\tau-x]\\&-\frac{1}{2}J_0\left[\frac{c}{\gamma}\sqrt{\gamma^2\tau^2-(x-\xi)^2}\right]\delta[x-(\xi-\gamma\tau)]H[\xi+\gamma\tau-x]\\&-\frac{1}{2}J_0\left[\frac{c}{\gamma}\sqrt{\gamma^2\tau^2-(x-\xi)^2}\right]H[x-(\xi-\gamma\tau)]\delta[\xi+\gamma\tau-x].\end{aligned} \quad (A.22)$$

In view of the substitution property of the delta function, the last two terms in (A.23) reduce to $-\frac{1}{2}\delta[x-(\xi-\gamma\tau)]-\frac{1}{2}\delta[x-(\xi+\gamma\tau)]$ since $J_0(0)=1$ and $H(2\gamma\tau)=1$.
Therefore we obtain

$$\begin{aligned}\int_{-\infty}^{\infty}f(x)\frac{\partial K}{\partial t}(x,0;\xi,\tau)dx=&-\frac{c\tau}{2}\int_{\xi-\gamma\tau}^{\xi+\gamma\tau}f(x)\frac{J_0'\left[\frac{c}{\gamma}\sqrt{\gamma^2\tau^2-(x-\xi)^2}\right]}{\sqrt{\gamma^2\tau^2-(x-\xi)^2}}dx\\&-\frac{1}{2}f(\xi-\gamma\tau)-\frac{1}{2}f(\xi+\gamma\tau)\end{aligned} \quad (A.23)$$

Combining these results and noting that $-J_0'(x)=J_1(x)$ the Bessel function of order one, gives the solution $u(x,t)$ of the initial value problem (A.4) as

$$u(x,t) = \frac{f(x-\gamma t)+f(x+\gamma t)}{2} + \frac{1}{2\gamma}\int_{x-\gamma t}^{x+\gamma t} g(\xi) J_0\left[\frac{c}{\gamma}\sqrt{\gamma^2\tau^2-(x-\xi)^2}\right] d\xi$$

$$-\frac{c\tau}{2}\int_{x-\gamma t}^{x+\gamma t} f(\xi)\frac{J_1\left[\frac{c}{\gamma}\sqrt{\gamma^2\tau^2-(x-\xi)^2}\right]}{\sqrt{\gamma^2\tau^2-(x-\xi)^2}} d\xi \tag{A.24}$$

$$+\frac{1}{2\gamma}\int_0^t \int_{x-\gamma(t-\tau)}^{x+\gamma(t-\tau)} F(\xi,\tau) J_0\left[\frac{c}{\gamma}\sqrt{\gamma^2(t-\tau)^2-(x-\xi)^2}\right] d\xi d\tau.$$

This solution formula reduces to that for the Cauchy problem for the inhomogenous wave equation if we set $c = 0$. Also if we set $c = i\hat{c}$ in (A.24) and note that $J_0(iz) = I_0(z)$ and $J_1(iz) = iI_1(z)$ we obtain as the solution Heaviside equation

$$u_{tt} - \gamma^2 u_{xx} + \hat{c}^2 u = F(x,t) \tag{A.25}$$

$$-\infty < x < \infty, \qquad t > 0$$

with the initial condition (A.4)

$$u(x,t) = \frac{f(x-\gamma t)+f(x+\gamma t)}{2} + \frac{1}{2\gamma}\int_{x-\gamma t}^{x+\gamma t} g(\xi) I_0\left[\frac{\hat{c}}{\gamma}\sqrt{\gamma^2 t^2-(x-\xi)^2}\right] d\xi$$

$$+\frac{\hat{c}t}{2}\int_{x-\gamma t}^{x+\gamma t} f(\xi)\frac{I_1\left[\frac{\hat{c}}{\gamma}\sqrt{\gamma^2 t^2-(x-\xi)^2}\right]}{\sqrt{\gamma^2 t^2-(x-\xi)^2}} d\xi \tag{A.26}$$

$$+\frac{1}{2\gamma}\int_0^t \int_{x-\gamma(t-\tau)}^{x+\gamma(t-\tau)} F(\xi,\tau) I_0\left[\frac{\hat{c}}{\gamma}\sqrt{\gamma^2(t-\tau)^2-(x-\xi)^2}\right] d\xi d\tau.$$

2. Ballistic Heat Transport in Graphene

Very important reason for the interest in graphene is a unique nature of its charge carriers. In condensed matter physics the Schrödinger equation rules the world, usually being quite sufficient to describe electronic properties of materials. Graphene is an exception: its charge carriers mimic relativistic particles and are easier and more natural to describe starting with the relativistic equations: Klein - Gordon and Dirac rather than the Schrödinger equation. Although there is nothing particularly relativistic about electrons moving around carbon atoms, their interaction with a periodic potential of graphene lattice gives rise to new quasiparticles that at low energies are accurately described by (2+1) dimensional equation with an effective speed of light $\upsilon \approx 10^6$ m/s. These quasiparticles, called massless Dirac fermions, can be seen as electrons that lost their rest mass m_0 or as neutrinos that acquired the electron charge [A.2 – A.4].

In the monograph [A.1] the quantum thermal equation for the heat transport was formulated.

$$\frac{1}{\upsilon^2}\frac{\partial^2 T}{\partial t^2}+\frac{m}{\hbar}\frac{\partial T}{\partial t}+\frac{2Vm}{\hbar^2}T-\frac{\partial^2 T}{\partial x^2}=F(x,t). \tag{A.27}$$

The solution of equation (A.27) can be written as

$$T(x,t)=e^{t/2\tau}u(x,t). \tag{A.28}$$

After substituting formula (A.28) into (A.27) we obtain new equation

$$\frac{1}{\upsilon^2}\frac{\partial^2 u}{\partial t^2}-\frac{\partial^2 u}{\partial x^2}+q^2u(x,t)=e^{t/2\tau}F(x,t), \tag{A.29}$$

and

$$q^2=\frac{2Vm}{\hbar^2}-\left(\frac{m\upsilon}{2\hbar}\right)^2. \tag{A.30}$$

Equation (A.29) can be written as

$$\frac{\partial^2 u}{\partial t^2}-\upsilon^2\frac{\partial^2 u}{\partial x^2}+q^2\upsilon^2u(x,t)=G(x,t), \tag{A.31}$$

where

$$G(x,t)=\upsilon^2e^{t/2\tau}F(x,t).$$

Equation (A.31) has the same form as the equation (A.1). As can be seen from Eq. (A.31) for $q^2=0$ we obtain ballistic heat transport with velocity $\upsilon=\alpha c\approx 10^6$ m/s as in graphene.

APPENDIX B. GENERALIZED MAXWELL-PROCA EQUATIONS FOR HEAVY PHOTONS

Electromagnetic phenomena in vacuum are characterized by two three dimensional vector fields, the electric and magnetic fields $\vec{E}(\vec{x},t)$ and $\vec{B}(\vec{x},t)$ which are subject to Maxwell's equation and which can also be thought of as the classical limit of the quantum mechanical description in terms of photons. The photon mass is ordinarily assumed to be exactly zero in Maxwell's electromagnetic field theory, which is based on gauge invariance. If gauge invariance is abandoned, a mass term can be added to the Lagrangian density for the electromagnetic field in a unique way:

$$L = -\frac{1}{4\mu_0} F_{\mu\nu} F^{\mu\nu} - j_\mu A_\mu + \frac{\mu_\gamma^2}{2\mu_0} A_\mu A^\mu, \tag{B.1}$$

where μ_γ^{-1} is a characteristic length associated with photon rest mass, A_μ and j_μ are the four - dimensional vector potential $\left(\vec{A},\ i\phi / c\right)$ and four – dimensional vector current density $\left(\vec{J},\ ic_\rho\right)$ with ϕ and $\vec{A}$ denoting the scalar and vector potential and ρ, $\vec{J}$ are the charge and current density, respectively, μ_0 is the permeability constant of free space and $F_{\mu\nu}$ is the antisymmetric field strength tensor. It is connected to the vector potential through

$$F_{\mu\nu} = \frac{\partial A_\nu}{\partial x_\mu} - \frac{\partial A_\mu}{\partial x_\nu}. \tag{B.2}$$

The variation of Lagrangian density with respect to A_μ yields the Proca equation (Proca 1930)

$$\frac{\partial F_{\mu\nu}}{\partial x_\nu} + \mu_\gamma^2 A_\mu = \mu_0 J_\mu. \tag{B.3}$$

Substituting Eq. (B.2) into (B.3) we obtain the wave equation of the Proca field

$$\left(\nabla^2 - \frac{\partial^2}{\partial (ct)^2} - \mu_\gamma^2\right) A_\mu = -\mu_0 J, \tag{B.4}$$

$$\left(\Box^2 - \mu_\gamma^2\right) A_\mu = -\mu_0 J. \tag{B.5}$$

In free space The *Proca* equation (B.5) reduces to (B.6), for a vector electromagnetic potential of $\vec{A}_\mu$.

$$\begin{aligned} &\left(\Box^2 + \mu_\gamma^2\right) A_\mu = 0, \\ &\Box^2 = \frac{1}{c^2}\frac{\partial^2}{\partial t^2} - \frac{\partial^2}{\partial x^2}, \end{aligned} \tag{B.6}$$

which is essentially the Klein – Gordon equation for massive photons. The parameter μ_γ can be interpreted as the photon rest mass m_γ with

$$m_\gamma = \frac{\mu_\gamma \hbar}{c}. \tag{B.7}$$

With this interpretation the characteristic scaling length μ_γ^{-1} becomes the reduced Compton wavelength of the photon interaction. An additional point is that static electric and magnetic fields would exhibit exponential dumping governed by the term $\exp\left(-\mu_\gamma^{-1} r\right)$ is the photon is massive instead of massless.

It is well-known that the electromagnetic constant c in Maxwell theory of electromagnetic waves propagating in vacuum and special relativity was developed as a consequence of the constancy of the speed of light. However, one of the prediction of massive photon electromagnetic theory is that there will be dispersion of the velocity of massive photon in vacuum.

The plane wave solution of the Proca equations without current is $A^\nu \sim \exp\left(ik^\mu x_\mu\right)$, where the wave vector $k^\mu = \left(\omega, \vec{k}\right)$ satisfies the relationship

$$k^2 c^2 = \omega^2 - \mu_\gamma^2 c^2. \tag{B.8}$$

As can be shown that

$$\upsilon_g \text{ (group velocity)} = c\left(1 - \frac{\mu_\gamma^2 c^2}{2\omega^2}\right) \tag{B.9}$$

and

$$\upsilon_g = 0 \quad \text{for} \quad \omega = \mu_\gamma c,$$

namely the massive waves do not propagate. When $\omega < \mu_\gamma c$, k becomes an imaginary quantity and the amplitude of a free massive wave would, therefore, be attenuated exponentially. Only when $\omega > \mu_\gamma c$ can the waves propagate in vacuum unattenuated. In the limit $\omega \to \infty$, the group velocity will approach the constant c for all phenomena.

A nonzero photon mass implies that the speed of light is not unique constant but is a function of frequency. In fact, the assumption of the constancy of speed of light is not necessary for the validity of special relativity, i.e. special relativity can instead be based on the existence of a unique limiting speed c to which speeds of all bodies tend when their energy becomes much larger then their mass. Then, the velocity that enters in the Lorentz transformation would simply be this limiting speed, not the speed of light.

It is quite interesting that the *Proca* type equation can be obtained for thermal phenomena.

APPENDIX C. GENEALOGY OF THE KLEIN – GORDON EQUATION

Heaviside Equation 1876

$$\frac{\partial w}{\partial t^2} + k\frac{\partial w}{\partial t} = a^2 \frac{\partial^2 w}{\partial x^2}$$

Klein – Gordon Equation
$$w(x,t) = \exp\left(-\tfrac{1}{2}kt\right)u(x,t)$$
$$\frac{\partial^2 u}{\partial t^2} = a^2 \frac{\partial^2 u}{\partial x^2} + \left(b + \frac{1}{4}k^2\right)u$$

physics, astrophysics, chemistry

Fisher Equation
$$\frac{\partial w}{\partial t} = \frac{\partial^2 w}{\partial x^2} + aw(1-w)$$
heat and mass transfer
biology, ecology, genetics

Schrödinger Equation
$$i\eta\frac{\partial \Psi}{\partial t} = -\frac{\eta}{2m}\Delta\Psi + V\Psi$$
physics, astrophysics, chemistry

FitzHugh – Nagumo Equation
$$\frac{\partial w}{\partial t} = \frac{\partial^2 w}{\partial x^2} - w(1-w)(a-w)$$
genetics, biology, medicine

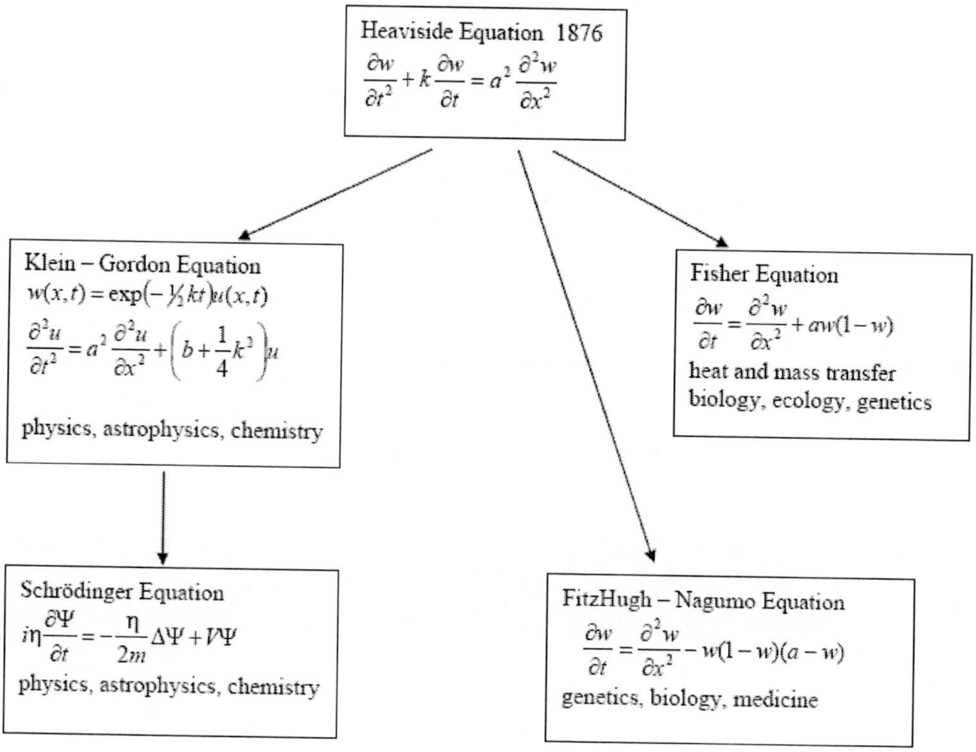

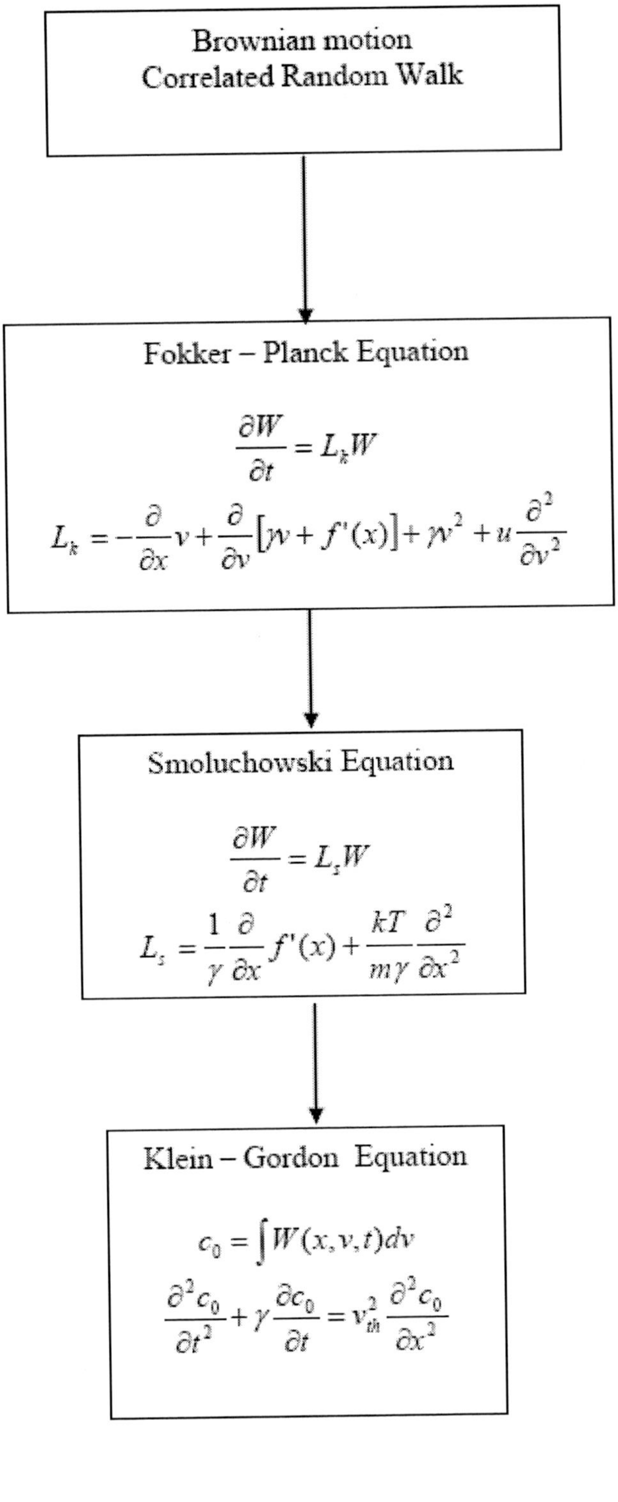
Brownian motion
Correlated Random Walk
Fokker – Planck Equation
$\frac{\partial W}{\partial t} = L_k W$
$L_k = -\frac{\partial}{\partial x} v + \frac{\partial}{\partial v}[\gamma v + f'(x)] + \gamma v^2 + u \frac{\partial^2}{\partial v^2}$
Smoluchowski Equation
$\frac{\partial W}{\partial t} = L_s W$
$L_s = \frac{1}{\gamma}\frac{\partial}{\partial x} f'(x) + \frac{kT}{m\gamma}\frac{\partial^2}{\partial x^2}$
Klein – Gordon Equation
$c_0 = \int W(x, v, t) dv$
$\frac{\partial^2 c_0}{\partial t^2} + \gamma \frac{\partial c_0}{\partial t} = v_{th}^2 \frac{\partial^2 c_0}{\partial x^2}$

Appendix D. HiPER- High Power Laser Energy Research Facility

HiPER is being designed to demonstrate the feasibility of laser driven fusion as a future energy source. It will also enable the investigation of the science of truly extreme conditions – accessing regimes which cannot be produced elsewhere on Earth (temperatures of hundreds of millions of degrees, billion atmosphere pressures, and enormous electric and magnetic fields). HiPER will require major developments in technology, building on the highly successful European capability in this area. In particular, the PETAL laser, located in the Aquitaine region of France, will be a fore-runner to the HiPER facility to address physics and technology issues of strategic relevance for HiPER Plentiful fuel (at a scale that can fully meet mankind's long term needs) Energy Security (sea water and the Earth's crust are the source of the fuel) Sufficiently Clean energy (there are no greenhouse gas emissions, and no long-lived radioactivity). Safe operation (there is no stored energy, so 'melt down' or catastrophic failure are impossible) HiPER offers a complementary solution to the fusion project known as ITER, which uses magnetic fields to contain the fusion reaction. The scale of the energy problem is such that multiple solutions are demanded. There is great potential for knowledge exchange between the two projects in areas such as material research, diagnostics and the underlying science. Fusion has been pursued for over 40 years, with lots of early mistakes with regard to anticipated timescales. Fortunately, both routes to fusion have received many billions of dollars investment, such that their future development path is now clear. The physics underlying inertial fusion is already proven. This is the approach adopted by Nature – this mechanism powers our Sun and all other stars. Far more importantly, the process of energy production from inertial fusion has already been demonstrated on Earth in a spin-out of the US defence program. Demonstration of net energy from inertial fusion using a laser is now anticipated in the period 2010 to 2012 (on the National Ignition Facility, USA). HiPER is designed to move forwards from this landmark demonstration, using an approach which better enables a commercial energy production solution (and a broad-based basic science mission). The principle is conceptually similar to a combustion engine – a fuel compression stage and an ignition stage. Lasers are used to compress a shell of Deuterium and Tritium fuel to very high density. A very high power laser is then focused into the dense DT fuel, raising it to fusion temperatures (~100 million degrees Celsius). The power of these lasers is truly immense: roughly ten thousand times the power in the entire UK national grid! Of course, this power only lasts for a few million millionths of a second and acts like a match to ignite the fusion fuel HiPER will open up entirely new research programmes in a wide range of scientific disciplines, including: Astrophysics in the laboratory, including studies of the physics associated with supernovae evolution, proto-stellar jets, planetary nebulae, interacting binary systems, cosmic ray seeding and acceleration, and gamma-ray bursters Behaviour of matter in truly extreme conditions (tens of millions of degrees temperature, pressures of billions of atmospheres, and magnetic fields a billion times stronger than that of the Earth) Warm Dense Matter studies – addressing the principal outstanding regime of material science in which there is no accepted theory (for which HiPER will offer exceptional probing and diagnostic capability). This is of direct relevance to planetary geophysics, tackling uncertainties in the evolution of Earth-like and giant gaseous planets. Turbulence – how do highly compressible, nonlinear flows transition to turbulence and subsequently evolve? This is one of the few remaining fundamental uncertainties in classical physics.

Laser-plasma interaction physics – including the question of how waves and matter interact under highly nonlinear conditions. Production and interaction of relativistic particle beams – for example, whether macroscopic amounts of relativistic matter can be created (then studied and utilized). Fundamental physics at the strong field limit – for example, studying the physics of the quantum vacuum, collective and dynamical QED studies such as photon propagation in the early universe, and possibly issues associated with Schwinger pair production and Unruh (Hawking) radiation.

Equations Index

INDEX

A

B

C

D

E

F

G

L

M

N

O

P

Q

R

S

T

U

V

W

X

Y